Off-Grid Solar Power Handbook

12 Volts Mobile Solar Power for RVs, Boats, Vans, Campers, Cabins and Tiny Homes

Andy Reed

Copyright © 2020 by Andy Reed
All rights reserved.
ISBN-13: 9798563313781

This document is geared towards providing exact and reliable information with regards to the topic and issue covered. The publication is sold with the idea that the publisher is not required to render accounting, officially permitted, or otherwise, qualified services. If advice is necessary, legal or professional, a practiced individual in the profession should be ordered.

- From a Declaration of Principles which was accepted and approved equally by a Committee of the American Bar Association and a Committee of Publishers and Associations.

In no way is it legal to reproduce, duplicate, or transmit any part of this document in either electronic means or in printed format. Recording of this publication is strictly prohibited and any storage of this document is not allowed unless with written permission from the publisher. All rights reserved.

The information provided herein is stated to be truthful and consistent, in that any liability, in terms of inattention or otherwise, by any usage or abuse of any policies, processes, or directions contained within is the solitary and utter responsibility of the recipient reader. Under no circumstances will any legal responsibility or blame be held against the publisher for any reparation, damages, or monetary loss due to the information herein, either directly or indirectly.

Respective authors own all copyrights not held by the publisher.

The information herein is offered for informational purposes solely, and is universal as so. The presentation of the information is without contract or any type of guarantee assurance.
The trademarks that are used are without any consent, and the publication of the trademark is without permission or backing by the trademark owner. All trademarks and brands within this book are for clarifying purposes only and are the owned by the owners themselves, not affiliated with this document.

Table of Contents

Warning
9

Introduction
11

PV System in 5 minutes
13

1. Battery
13
2. Solar Panel
14
3. Charge controller
14
4. Inverter
14
5. Wires
15

LEARNING
16

Basic Electricity Notions
17

Understanding an Electrical circuit
17

AC/DC
18

Units
20

Useful Formulas … 23

Basic circuits … 26

Batteries … **33**

Starting Batteries … 33

Deep Cycle Batteries … 34

Dual Purpose Batteries … 34

And the Winner is … 35

Choosing your Deep Cycle Marine battery … 35

Your Battery's Assistant … 44

Multiple Batteries … 45

Pairing with solar … 49

Should You Buy Used Batteries? … 50

Solar Panels … **52**

Mind the Voltage … 52

Multiple Solar Panels … 54

Choosing Your Solar Panels … 59

Get the Most Out of Your Panels … 67

Should You Buy Used Panels? … 77

Charge Controller — 79
PWM and MPPT — 81
Selecting a Charge Controller — 82

Inverter — 88
Best Inverters for Off-Grid Applications — 88
Output Signal Types — 90
Selecting an Inverter — 93

Wires and Stuff — 95
The Right Wire Type — 95
The Right Wire Size — 99
The Right Wire for the Right Use — 101
Fuses — 104
Busbars — 105
Shunts — 105
Circuit Breakers and Isolator Switches — 106

National Electrical Code (NEC) — 109

DOING — 110

Designing the System — 111

Sizing Your Needs — 111

Calculating Battery storage — 114

Calculating Solar Power Output — 124

Planning the locations — 127

Choosing the Charge Controller and the Inverter — 129

Sizing Wires — 131

Sizing Fuses — 137

Notes on Appliance Efficiency — 138

Final System Design — 140

Setup — 147

Tools and Equipment — 147

Setup Procedure — 151

Maintenance and Storage — 161

Conclusion — 165

Will you do Me a Favor? — 167

Warning

The content in this guidebook is fashioned in simple terms. For more detailed information, it is best to check with a qualified electrician.

Electricity is a dangerous thing, and you are not supposed to mess with AC household electricity. In fact, it's illegal.

However, you can work with DC current, which is what we are going to use here, especially at lower Voltage, but it still dangerous if a circuit is not constructed properly.

If you have any doubts about performing DIY electrical work, please hire a qualified electrician to perform the job. No one will blame you for not being "smart enough" or any other nasty, stupid thing. In fact, rather than being a "failure," having read this book, having an electrician to check if you are doing right, or helping you out with the setup and following the National Electrical Code may be the smartest thing to do.

In fact, that is precisely what I've done in the beginning, and here I am.

Now, let's get started!

Introduction

This book is meant for people like you and me who love learning new stuff and getting their hands dirty.

When I first approached the world of Solar Panels and photovoltaic systems, I was overwhelmed by the amount of information out there, so I reached out to my friend Jim, a professional electrician, to cut through the noise and go straight to the point.

With Jim's help, I've been able to get my first off-grid PV system up and running at my cabin. I kept digging, and I designed and installed another PV system on my Camper Van that works wonders.

Since then, I've helped tens of friends to mount off-grid solar, and if you're reading these very pages, it's because one of them encouraged me to share my knowledge with the world (thank you, Stan!)

This guide collects all the essential information - both practical and theoretical - that I've learned through the years. They result from intense studying, uncountable trials, errors, and many beers offered to my friend Jim.

In this book, I will provide the exact information that you need to do the job. I'll cut off all the unnecessary stuff that other authors use to fill their books and increase their page count.

After reading this book, you should be able to mount Your basic PV system in less than a day. And you can trust me, it will give you a deep sense of achievement!

Most of us first took this PV path to reduce their energy bills or to make sure they have a reliable energy supply on their camper truck. However, when you'll finish installing the whole system and start getting energy out of it wherever you are, you will experience a sense of pure freedom. And this, my friend, is priceless.

PV System in 5 Minutes

If you are reading this book, it's because you are willing to learn how to build a basic 12 volt PV system, so I'll only provide the information that is absolutely necessary to get it done in the shortest time possible.

However, before digging into the details, I always find having a generic overview of the job to do is the best way to start planning out the work.

Thus, in this chapter, we will cover the 5 fundamental elements of our PV system. Let's go!

1. Battery

When talking about off-grid systems, the Battery is the most critical component you'll have to worry about. Even more than Solar Panels themselves. To be accessible when you need it, the energy you produce has to be accumulated, and that's where batteries come in. In the next chapters, you'll learn how to choose the right Battery and how to get the most out of it.

2. Solar Panel

Well, not much to say here. Why Solar Panels are critical to a PV system is pretty obvious, right?

3. Charge controller

The power generated by your PV panel will continuously flow to your Battery, even when it is 100% charged. This is called overcharging and can irreversibly damage your Battery in just hours. That's why you'll need a charge controller to monitor the State of Charge (SOC) of the Battery, regulate the Current flow, and stop it when full-charge is attained.

4. Inverter

Solar Panels and batteries operate with DC signal, while most loads out there use AC signal (we'll cover the difference between DC and AC in a moment). Therefore you'll need an inverter to convert the signal and get your solar power to work with your favorite devices.

5. Wires

Last but not least, the wires. There is no electric system without wiring. To build up your PV system, you'll need wires and everything that comes with it: fuses, wire lugs, crimp connectors, and so on.

Section 1
LEARNING

Basic Electricity Notions

Let's get friends with the generic notions and terms of the electric realm!

Understanding an Electrical circuit

In order to reach your devices and appliances, the energy you draw from your solar panel has to travel through an electrical circuit. Let's see how this works.

Basically, electrical circuits are just like racing circuits. Cars fire their engines at the starting line and run at high speed all along the circuit until they end up at the same spot where they began.

The same goes for electricity. The starting point is your battery's positive pole (red). The electrons race through the copper wire until they reach the battery again, this time on the negative terminal (black).

Now, what would happen if the racing car doesn't have pit stops along the circuit? Tires will melt, and the whole vehicle put at risk.

Again, the same goes for electricity. If you don't put "pit-stops" or barriers along the circuit (switches, fuses, fridges, lights, etc.), the

Current will circle non-stop at high speed through the wires, causing overheating and possibly bursting the whole system.

Good news is, a similar endless loop, which is known as a "closed circuit," is utterly useless to us. We want to power up our devices and appliances. That's the whole point of it.

So this is what we do. We place our obstacles on the circuit to use up electricity, and that's how we light our houses and watch Tv on the couch. And as long as you do this, your circuit should be fine.

But remember that in the end electricity has to come to the finish line. Otherwise, nothing will work.

Another interesting aspect that is worth noting is that electricity won't stop flowing. A lightbulb without a circuit breaker will stay on forever. Or at least until the bulb doesn't get consumed or the battery doesn't exhale its last breath. That's why we use switches to break the circuit and stop the electricity from flowing.

AC/DC

You've probably known that electricity flows in two ways, either AC or DC, since forever. But what does this mean concretely?

DC stands for Direct Current, and it describes an electrical signal that flows across the circuit steadily in a single direction.

AC stands for Alternating Current, and it describes an electrical signal that keeps switching directions.

AC is used to deliver power to our homes, offices, and industries because of its many advantages over DC when it comes to transmitting electricity over large distances without too much energy losses.

Very nice Andy, but what about real life?

Well, I will oversimplify this a bit and make some generic statements here that are not always true but are mostly true.

As I mentioned, AC is used to transport electricity over the grid, so the classic signal that comes to your house's plugs is AC.

The electricity generated from motors, dynamos, and similar is usually AC.

The electricity generated from batteries, Solar Panels, and similar is usually DC.

Heavy appliances like fridges, conditioners, washing machines, and similar use motors, pumps, and compressors on the inside and thus operate with AC.

Electronic devices like laptops, TVs, audio sets, and similar, on the other hand, operate with DC. They usually have an internal or external inverter that turns the AC signal from the wall in DC.

As I said, this is generic information. Always check your devices and appliances before messing with them. AC and DC don't quite like each other, and if you force them to talk by plugging AC in a DC plug or vice versa, this can end very badly for your device.

Units

Volt - Voltage

Voltage is the difference in charge between two points. That difference is what moves electrons through a circuit generating electricity.

You can think of it as the amount of pressure required to move the electrons.

In the case of your battery, Voltage is the difference in charge, or electric potential, between the positive pole and the negative pole.

Voltage is measured in Volts (V).

Conventionally, Voltage is indicated with the letter V.

Amp - Current

Current is the rate at which charge is flowing through the circuit.

Current is measured in Amperes or Amps (A), but to better deal with the numbers, Current is often expressed in milli-amps.

Conventionally, Current is indicated with the letter I.

Ohm - Resistance

Resistance is a material's tendency to resist the flow of charge (or Current).

A resistor will create friction and slow down the course of electrons, therefore reducing the rate of the flow.

Every material as a different resistance value. Copper, for example, has a very low resistivity, which makes for a great conductor. That's why it is commonly used to make wires and allow Current to flow easily across a circuit.

On the contrary, carbon has a high resistivity, which is why it is usually used in resistors.

Resistance is measured in Ohms (Ω).

Conventionally, resistance is indicated with the letter R.

Watt - Power

Power can be defined as the amount of energy consumed to do a specific task.

A standard incandescent lightbulb, for example, consumes 60 Watts of energy.

Power is measured in Watts (W).

Conventionally, power is indicated with the letter P.

Electrical Energy consumption

To calculate how much energy is consumed by our appliances, we usually consider either how much power they draw in an hour (Watt-hours or Wh) or how much Current they need in an hour (Amp-hour or Ah).

When consumption is high, it is usually expressed in kiloWatt-hours (kWh).

Amp-hours are commonly used to indicate the amount of energy batteries stores, so you'll want to keep an eye on this Ah sign..

Useful Formulas

Learning to calculate the energy consumption of our appliances is key to design a PV system that covers our needs. That is why we will have to learn some basic formulas. Don't worry, nothing difficult here, I promise. In fact, you'll be surprised by how easy it is.

Electrical variables symbols

- Voltage = V *(unit being Volts)*
- Current = I *(unit being Amps)*
- Resistance = R *(unit being Ohms)*
- Power = P *(unit being Watts)*
- Time = t

Ohm's Law

Current, Voltage, and Resistance are closely related to each other, and that relation is described by Ohm's law.

```
V = I × R
I = V ÷ R
R = V ÷ I
```

Power

```
P = V × I
V = P ÷ I
I = P ÷ V
```

Amp-hours

As we already discussed, Amp-hours is the unit used to calculate the capacity of battery banks, which is a critical value to consider when designing your PV system.

```
Amp hours = Amps × hours
```

This formula will give you an idea of how much Current you can draw from the battery in an hour to power up your appliances.

For example, a basic AA alkaline battery with a capacity of 2 Ah will supply 2 Amps in one hour.

It could also supply 1 Amp in two hours and so on. To find out the Current consumption in a different time frame, you just have to divide Amp-hours by the time (expressed in hours) that you're looking for.

```
Ah ÷ t (in hours) = Amps × your time frame
```

Examples:

How much Current would a 10 Ah battery provide in 5 hours?

```
10 Ah ÷ 5 h = 2 Amps in 5 hours
```

How much Current would a 10 Ah battery provide in half an hour?

```
10 Ah ÷ 0.5 h = 20 Amps in half an hour
```

Energy or Watt-hours

This unit is used by electrical companies to bill your energy consumption.

You may be used to kiloWatts-hours, though. That's because Watt-hours is a large number. Therefore, kiloWatt-hours is more convenient to use.

```
1 kWh = 1,000 Watt
```

Therefore, to calculate the amount of energy you consume in one hour, you use the formula:

```
E = P × t
```

Example:

```
E = 2,000 W × 1 hour

E = 2,000 Wh    or    2 kWh
```

In order to know how much money your energy needs cost you, you'll have to check the rate per kWh of your energy provider. In the USA, the average rate is $0.12 per kWh.

When you do such calculations, remember to convert Watt-hours in kiloWatt-hours.

Basic circuits

Basic Electrical Symbols

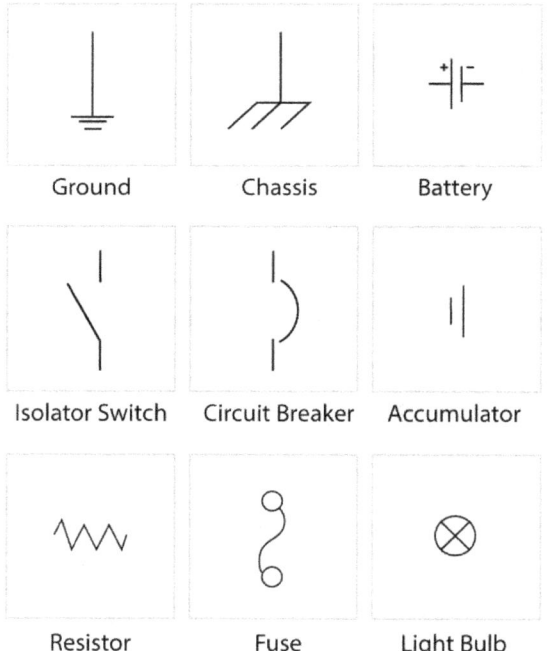

Ultra Simple circuit

This is the scheme for the basic circuit. You've got the power source, the switch, and the light bulb.

Remember red wire for positive, black wire for negative. In the following diagrams though, we'll be using blue for negative, just for visual clarity's sake. Black will display neutral.

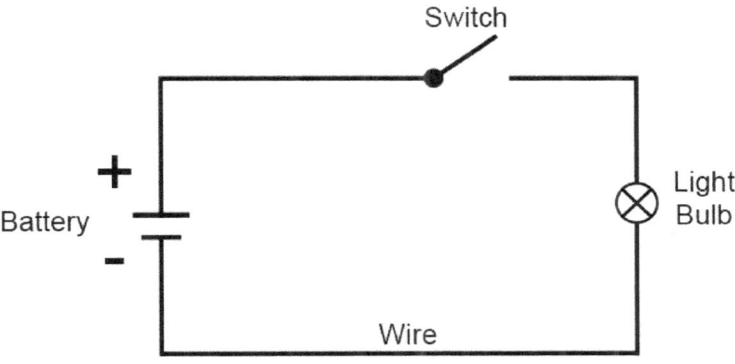

Series and Parallel

If you need multiple light bulbs (or appliances), you have two options. Either you mount them in series or in parallel.

Here's the difference.

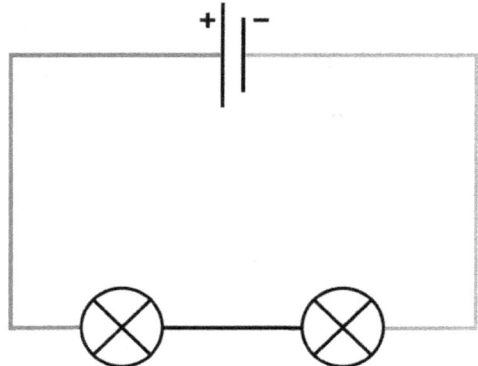

In a series circuit, there is only one path for Current to flow. This means that starting from the positive terminal (+) of the battery, electricity will first pass through Light 1, then through Light 2, and finally through Light 3 before starting off again.

The thing is, each light (and each appliance) has a resistance that slows down the Current flow. Therefore, there is less Current going through, and the lights are dimmer.

That's because of Ohm's law. The same Voltage divided by more resistance equals less Current.

Another downside is that a switch on a series circuit will open, or break, the entire thing and turn all the lights off.

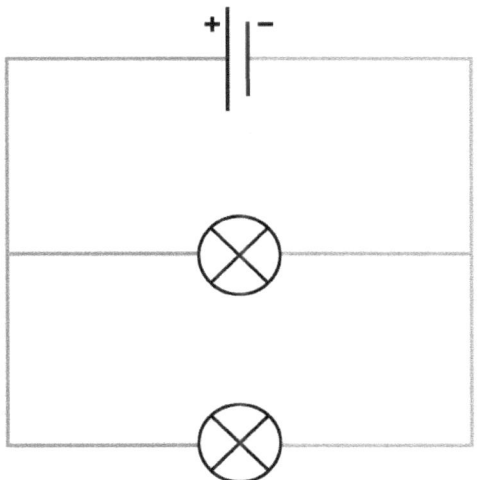

In a parallel circuit, however, Current has multiple paths to flow. Each individual path is called a branch and is, in fact, a different loop. Each branch has its own resistance, and is not affected by the resistance insisting on other branches.

Therefore, each bulb gets the full amount of Current and lights properly. You can also install a switch on every branch to have control over single lightbulbs.

Of course, you can also design a circuit that is a combination of series and parallel. In this example, I've added an extra switch in series so that I can turn all the lights off at once if I want to.

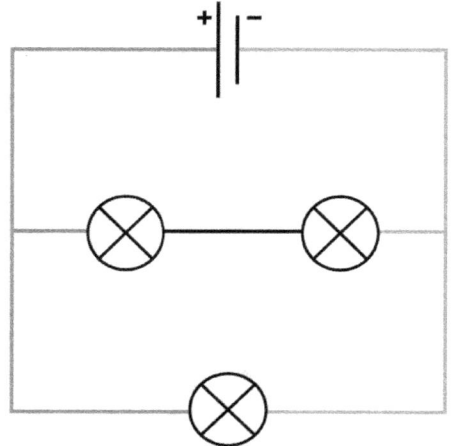

Safety: Fuses

Now that we've seen how basic circuits work, let's talk about safety for a minute.

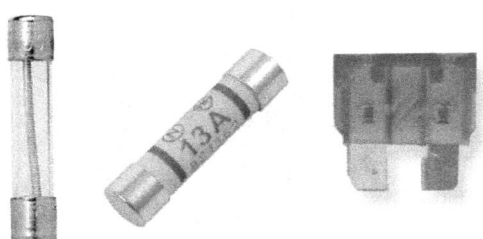

Sometimes for many reasons, an excessive amount of Current can occur, causing short circuits, overloading, mismatched loads, or device failure that can damage your devices.

To avoid that, we use fuses.

Fuses are basically small components that hold metal wire designed to melt when too much Current flows through it. When the fuse fuses, the circuit breaks, this prevents that excessive amount of Current from ruining the rest of the system. Long live fuses!

Safety: Grounding and Earthing

Another fundamental thing to ensure safety is grounding.

As the name suggests, Grounding (sometimes called Electrical Ground) is an electrical connection to the ground. Why do we do that?

Because high-Voltage electricity that can occur during power surges is dangerous. It can instantly fry your appliances, possibly start a fire and even cause physical injuries.

As Earth is the best conductor and electricity always chooses the least resistive route, Grounding gives excessive Current somewhere to go other than into you, possibly saving your life. It is like a backup pathway in case there's a fault in the wiring system, discharging the fault Current to the ground and restoring balance.

To ground an electrical system, you need to place a wire between the Current carrying part (neutral wire) and Earth via a copper ground rod.

Earthing (sometimes called Chassis Ground) is the same thing, the only difference being you'll place a wire between the equipment body (like your solar panel) and Earth. Earthing is usually made with green wires.

In a mobile setting like RVs, "ground" is the chassis of the vehicle.

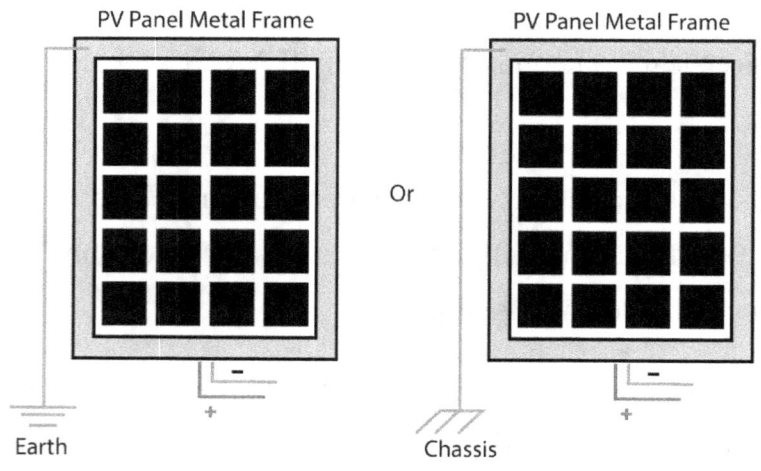

Batteries

In this chapter, we will learn to choose the right battery for your off-grid solar power system.

Batteries are basically energy accumulators. They store electricity that was previously produced, and they are able to deliver it when it is actually needed. That's exactly what you need since Solar Panels can't produce energy on demand and are highly dependent on weather conditions.

So let's have a look at the types of battery we can choose from.

First-off all, Batteries are NOT born equal. Understanding this simple concept is of the uttermost importance here.

Different batteries serve different purposes. There mainly are three types of batteries:

Starting Batteries

Their task is to start engines. Therefore they are really good at firing up huge amounts of power in a really short time. This type of batteries needs higher cranking amps or the ability to discharge and sustain a huge amount of power for 30 seconds at a given temperature (that's why cold-cranking amps also exist).

Of course, this is not the kind of battery you'll want to use to sustain your system.

Deep Cycle Batteries

Contrary to starting ones, deep cycle batteries are designed to support a steady energy supply over a long time. This feature makes deep cycle batteries ideal for running electrical appliances, and that's why they are used on leisure boats. Deep cycle batteries are also able to withstand many more cycles of charging and discharging.

Dual Purpose Batteries

This type can be used, as its name suggests, to both deliver the initial energy boost and support the long-term usage of appliances. As you can imagine, they are pretty amazing if you need both functions but can be very limiting if you only need one.

And the Winner is

Car batteries are cheaper to buy, so many will be tempted to build their PV system around this accumulator type. As you may have figured, though, car batteries are starting batteries, and therefore they are not suited for the job.

On the other hand, deep cycle marine batteries are just perfect for doing what we're looking for here, that is, sustain the usage of lights, laptops, a fridge, etc. They may cost a little more than car batteries. Still, they will save you money in the end, as car batteries will deteriorate quickly and will not be able to properly keep everything up and running.

Therefore, a deep cycle marine battery, also known as a leisure battery, should be your battery of choice for your off-grid system.

Choosing your Deep Cycle Marine battery

So far, so good. We've identified the type of battery we're going to need. Now it's time to have a look at the different specs you need to consider when choosing which battery to buy.

Technology: Lead-Acid vs Lithium

When it comes to Batteries, there are mainly two technologies out there: Lead-Acid or Lithium.

As you will find out, Lithium batteries are way better than Lead-Acid, so theoretically, there's no question. However, Lithium batteries are much more expensive than Lead-Acid, so we'll cover both here in order to give you a complete overview.

Depth of Discharge

Batteries don't like to be fully discharged. Completely draining them will cause accelerated deterioration, and they'll start to provide less and less power until they pass out completely, way sooner than you'd expect.

That's because batteries work through electrochemical processes that convert chemical energy into electrical energy and back. To enable these processes, there must be a certain balance between the chemical elements inside the battery. If this balance falls off, chemical elements deplete, causing unwanted reactions that affect the energy conversion's proper functioning.

In simple terms: over-discharging shortens the lifespan of your battery significantly. So don't drain it.

Different technologies have different thresholds, though.

For Lead-Acid Batteries, you don't want to discharge below 50%. This means that even if the battery says 100 Ah, you can only use 50 Ah.

Lithium batteries, on the other hand, can be discharged to 20%. This means that for a 100 Ah battery, you can use up to 80 Ah.

Number of Cycles

The action of charging and discharging a battery is called a cycle. The number of maximum cycles of a battery indicates how many times you'll be able to charge and discharge before the battery is dead.

If the recommended depth of discharge (DOD) is respected, Lead-Acid batteries can do as much as 1,200 cycles on average.

From Lithium Batteries, you can get up to 5,000 cycles, depending on the manufacturer.

Efficiency

Different batteries also have different efficiency when it comes to charge and discharge.

Long story short, Lithium has a 95-99% efficiency while Lead-Acid has an average of 80% (some can get to 95%, though).

Charging and discharging batteries is not free. You don't just feed them with 10 Volts and get 10 V. There's a part of the energy produced that goes as waste heat, and that means that you lose energy every time you make a charge cycle. If you remember Ohm's law, this makes perfect sense since a battery has some kind of resistance. As internal resistance increases (you drawing too much power at one time), battery efficiency decreases.

Capacity and Discharge rates are here to help.

Capacity is the actual Amp-hours available for you to use at the rated Voltage.

The discharge rate is the number of Amps you are actually drawing at a given time.

If you use a lot of electricity in a short amount of time, the discharge rate is higher. Therefore the capacity is reduced.

Every battery has an ideal discharge rate, expressed as a C-rate, or the maximum current the battery can actually safely deliver, so when choosing a battery, you'd want to check it.

Here's a reference table to show you how it works.

C-RATE	TIME
10C	6 min
5C	12 min
2C	30 min
1C	1h
0.5C or C/2	2h
0.2C or C/5	5h

That's a lot of things to consider, I know, lucky enough most manufacturers provide us the recommended discharge rate in the form of hours (Typically 20 HR).

Once you've got this discharge rate (let's keep the 20 HR here), you just have to use it to divide the battery's size.

Example:

```
200 Ah
20 HR
200 Ah ÷ 20 HR = 10 A
```

Therefore, if you draw 10 A constantly, your battery should last 20 hours.

In fact, the manufacturer is telling you that in 20 hours, your battery will go from 100% to 0%, which you should never do, as we previously said.

Of course, with the Solar Panels ideally supplying constant power to your Batteries, this should never be a problem.

Temperature

Another aspect to consider is the external temperature that will affect the battery.

Battery performance is highly dependent on both cold and heat. Batteries are usually designed to operate at 68°F (20°C), so every temperature variation will have an impact on its functioning.

However, cold and heat have different effects on the chemicals inside the battery cells.

Cold will slow down the processes, increasing resistance, and therefore reducing the actual capacity or Amp-hours available for you to use.

The good news is, at least, the lifespan of a similar slowly operating battery gets extended.

Heat has the exact opposite effect. External high temperatures allow electrons to move quicker, therefore reducing the resistance and increasing the overall capacity.

Bad news is, the lifespan of a battery operating in hot temperatures for a more extended period of time will dramatically drop. Heat is the worst enemy of batteries.

There also are slight differences regarding what the battery is doing at extreme temperatures. If it charges or if it discharges.

I don't want to go technical here, so let's keep it simple.

Lead Acid batteries can be charged and discharged from -4°F (-20°C) to 122°F (50°C). However, you should not discharge it below freezing, as the electrolytes can freeze and permanently damage your battery. From a charging point of view, you can still do it below freezing, but only if you reduce the C-Rate to 0.3 C (3 hours). When hot, you should lower the Volt threshold by 3mV per °C.

Lithium batteries can be discharged from -4°F (-20°C) to 140°F (65°C), but when it comes to charging, the range shrinks from 32°F (0°C) to 113°F (45°C). No charge is permitted below freezing, and even if performance is still good at higher temperatures, modern chargers prohibit charging above 122°F (50°C).

To avoid being excessively affected by temperature conditions, batteries should be placed on the inside, where you can keep them at room temperature.

Be careful though, they need sufficient room and venting to breathe in order to disperse heat. Otherwise, even if the room temperature is good, the temperature in the battery spot can rise way above the recommended levels.

Weight

The weight of batteries is another variable to consider when choosing the batteries.

It is crucial that you carefully plan where your batteries are going to be located before buying them. Especially if you are building up a system on an RV.

You can find the exact weight information in the manufacturer's datasheet. Still, generally speaking, you can keep in mind that Lithium batteries are two or three times lighter than Lead-Acid ones.

Safety

Lithium batteries have no safety issues. However, Lead-Acid, due to the electrochemical processes occurring on the inside, emit gasses like oxygen and hydrogen, which, if not taken care of, can be dangerous and even lead to an internal explosion. This will cause the spilling of acid and the case to burst.

I don't want to scare the hell out of you, many many people use Lead-Acid batteries, and everything goes just fine.

This is just to warn you and give advice on the different types of Lead-Acid batteries and how to maintain them to avoid any safety issue.

Lead-Acid divides into two types:

- Vented Lead-Acid (VLA or Flooded Lead-Acid)
- Valve Regulated Lead-Acid (VLRA or Sealed Lead-Acid)

VLA batteries have a small removable tap on the top that gives access to their internal structure. The technology behind them causes the expulsion of gases due to electrolyzes. Therefore placing them in well-ventilated areas to prevent these gasses from accumulating is critical to ensure safety.

From a maintenance point of view, you'll also need to refill VLA batteries with distilled water, as the electrolyzes cause water evaporation. This little operation is usually only necessary every three months. You'll also need to clean the terminals to remove oxide.

VLRA batteries, instead, are sealed, meaning that you can't access the interior of the battery. They don't need a water refill and have a reduced need for well-ventilated areas. They still need it, though, because VLRAs are equipped with a pressure sensitive-valve that can open in case something's wrong inside the Battery to release exceeding gases.

VLRAs can be Gel or AGM. Both have the same safety advantages since they both have a system to reincorporate evaporated water back in the cells instead of releasing it. Both are also safer than VLA as they use silicon (Gel) or fiberglass (AGM) to prevent sulfuric acid from spilling out if the case breaks.

Your Battery's Assistant

Batteries are just batteries. They do their job the best they can, taking electricity in and delivering it when you need, but they don't know how to regulate the "flow."

Even if they are 100% charged, if Current keeps coming, they will try to bring it in (causing overheating, thermal runaway, and decomposition).

Same goes for discharge. Even when they reach the Depth of Discharge threshold (50% for Lead-Acid, 20% for Lithium), Batteries will stubbornly do what they know better: serve you with the electricity you demand.

They just don't know when to stop.

This is why we use Charge Controllers to regulate the "flow" and disconnect the battery when it reaches a 100% charge or the DOD threshold. Overcharging and over-discharging problems are solved. Yay!

To deal with over-discharging, you can use a low-Voltage disconnect system (automatic disconnection) or a shunt (manual disconnection), too.

We'll have a more in-depth look at Charge Controllers later on.

Multiple Batteries

You may find that the Amp-hours of one battery are not enough to meet your electricity needs.

No worries, my friend, you can add as many batteries you want to your PV system.

To use multiple batteries at once, you can choose from three configurations that should remind you of something: series, parallel, or a combination of both.

What's the difference? Well, it works just like in our lightbulb example in the previous Basic Circuits chapter. The only difference is that batteries are a power source. They don't use Amps. They provide Amp-hours.

Series

When connected in series, batteries get a higher Voltage (24 V, 48 V) but stick to the same capacity (Amp-hours).

Batteries in Series

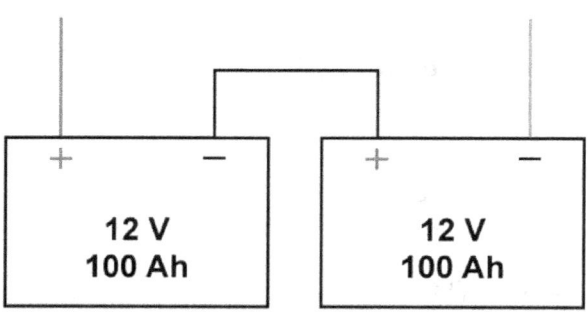

Total Voltage: 12 × 2 = 24 V
Total Amp-Hours: 100 Ah

Basically, when you connect batteries in series, you sum up their Voltage. Two batteries of 12 V result in a 24 V power source, three batteries are 36 V, four batteries are 48 V, and so on.

Since most charge controllers are only able to operate with 12 V, 24 V, and 48 V configurations, I recommend choosing one of these Voltage options.

However, 9 out of 10, you'll be looking for more Amp-hours rather than more Voltage. Therefore this is not a good solution if you need to use more appliances altogether.

NB: Resistance increases when batteries are connected in series.

Parallel

When connected in parallel, batteries get a higher capacity (Amp-hours) but stick to the same Voltage (12 V).

Batteries in Parallel

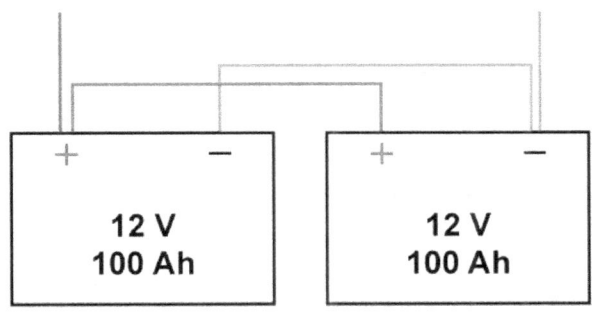

Total Voltage: 12 V
Total Amp-Hours: 100 × 2 = 200 Ah

Basically, when you connect batteries in parallel, you sum up their capacity (Amp-hours). Two batteries with a 110 Ah rate result in a 220 Ah capacity, three batteries are 330 Ah, four batteries are 440 Ah, and so on.

When you connect batteries in parallel, though, every battery should have the same Voltage and capacity, so don't mess with Volts and Amp-hours, and everything should be fine.

This is the ideal solution if you need more hours of power.

NB: Resistance decreases when batteries are connected in parallel.

Attention, please! Some Lithium batteries are not suited for parallel connection. The parallel connection for Lithium is entirely dependent on the built-in BMS (Battery Management System). You may also be able to buy and install an external BMS, but you should always check with the manufacturer.

Series + Parallel

A combined system with both series and parallel connection provides higher Voltage as well as higher capacity.

Here's how you can achieve this.

Batteries in Series and Parallel

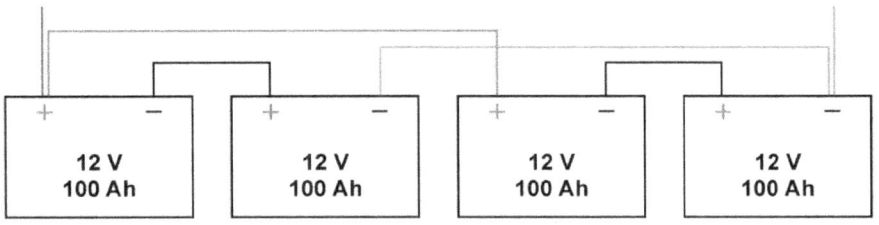

Total Voltage: 12 × 2 = 24 V
Total Amp-Hours: 100 × 2 = 200 Ah

If you are building a standard off-grid PV system, it is unlikely that you'll need more than a 12 V Voltage. However, for more demanding applications (500-2,000 W), a 24 V may be best suited. 48 V configuration should only be used for higher power configurations.

Since this book is about 12 V systems, we'll stick to this kind of configuration from now on.

Pairing with solar

Solar Panels suited for 12 V batteries usually have a Voltage of around 17 V. This is not a problem for our PV system since the Charge Controller will take care of adjusting the Voltage to enable a better "communication" between the panel and the battery.

So why am I even telling you this?

Because you may want or need to connect two or more batteries to your system to store more electricity and multiply the Amp-hours available (we will get there in a minute). However, suppose you mount two batteries to have 24 V (12+12). In that case, a single Solar Panel producing 17 V will not be able to power them. So be sure to consider this information when you are thinking of expanding your power storage capacity.

Spoiler: as for batteries, you can also join Solar Panels together, in series or in parallel, depending on your need to get extra power.

Should You Buy Used Batteries?

In this chapter, you've learned how delicate batteries are. How the way they are being used, the external temperature, and many other factors can influence their efficiency and lifespan.

Therefore, buying used batteries, even if it's a good deal, is not a good idea.

You don't know how many cycles are left, nor you know how much the battery has been left self-discharging somewhere in a cold or hot environment. A perfectly good-looking battery case can hide a terribly deteriorated battery that can last only a few months.

Solar Panels

Ok, so we've learned how to choose batteries to store electricity. I guess it's now the time to dig into how we are going to produce this electricity in the first place.

Let me introduce you the star of this book! I'm talking about... Solar Panels!!!

Whether you want to build a PV system for your camper, caravan, boat, cabin, or whatever, Solar Panels are the ones that will harvest sunlight for you, turning it into electricity for you to refrigerate your food, light your RV at night, and even watch Netflix.

So first of all, let's thank Mr. Solar Panel for making our life easier. Thank you, my friend.

Alright, let's get down to it.

Mind the Voltage

In any off-the-grid system, Solar Panels are, actually, the grid, as they provide electricity to your system.

Having a Solar Panel constantly connected to your battery is just like having your smartphone constantly plugged into the wall: you've got an endless power source that top-up the battery your device is relying on.

Marvelous right? Well, yes, but only if certain conditions are met:

- Sunlight must hit the panels
- Solar Panels have to provide enough Voltage to top-up the battery

The first one is easy. Solar Panels only work with solar radiation, aka sunlight. By night they completely stop generating electricity. By day their production is affected by the weather. If the sky is blue: high production. If it gets cloudy: far less production. Same goes when the panels get shaded.

This is why you have a battery installed.

As for the Voltage issue, here's how to deal with it. For the energy to flow from the panels to the battery, the panels should have a higher Voltage than the battery. In general, the power source should always have a higher Voltage. It's as simple as that.

This means that, if you have a 12 V battery, your Panels should provide more than 12V (at least 13.6 V). Otherwise, it won't charge anything.

Typically, average Solar Panels are rated around 17 V, so you should be fine with one 12 V battery. But if you have multiple 12 V batteries connected in series (which means higher Voltage), 17 V will not be enough.

Multiple Solar Panels

One Solar Panel is usually not enough to provide enough energy for our needs. Thus, you'll likely need to join at least two panels together to get enough production to sustain all of your appliances.

Joining Solar Panels together is just like joining batteries. You can join them in series, in Parallel, or with a combination of both.

Series

When connected in series, 17 V Solar Panels reach a higher Voltage (34 V, 51 V), but the Amperage (Current) remains the same.

Basically, when you wire Solar Panels in series, Voltage adds-up. Two panels of 17 V result in a 34 V power source, three panels are 51 V, four panels are 68 V, and so on.

As we mentioned in the Multiple Batteries chapter, most Charge Controllers are only able to operate with 12 V, 24 V, and 48 V configurations. Therefore it is recommended to choose one of these

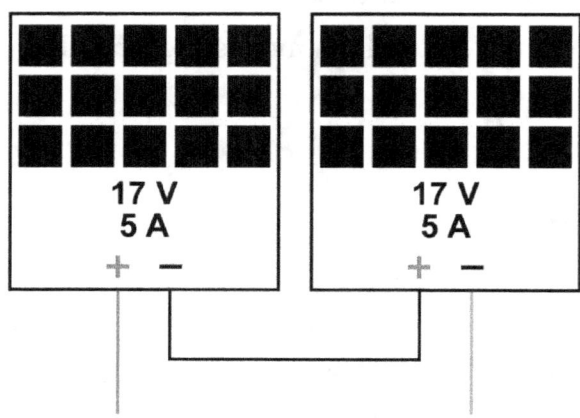

Total Voltage: 17 × 2 = 34 V
Total Amps: 5 A

Voltage options. Solutions exist for higher Voltage needs, but here we are going to stick to the basics of a 12 V system. I just want to raise awareness on the importance of choosing the right Charge Controller for your specific Voltage needs.

Also, if you connect Solar Panels in series, an MPPT Charge Controller is needed. More on that in the specific chapter.

A major drawback for series wiring is that if one panel is defective, the whole circuit fails.

As always, when it comes to joining PV panels, the best choice is to connect panels with the same characteristics (Voltage and Amperage) and possibly from the same manufacturer. If you wire Solar Panels with different specs in series, the Voltage will add-up as before, but the Amperage output will align to the lowest panel.

Parallel

When connected in Parallel, 17 V Solar Panels maintain the same Voltage, but Amperage (Current) increases.

Basically, when you wire Solar Panels in series, Amperage adds-up. Two panels of 5 A result in a 10 A power source, three panels are 15 A, four panels are 20 A, and so on.

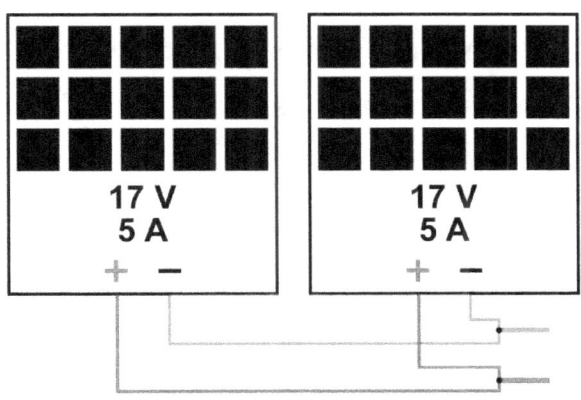

Solar Panels in Parallel

Total Voltage: 17 V
Total Amps: 5 × 2 = 10 A

When you connect batteries in Parallel, though, every battery should have the same Voltage and capacity, so don't mess with Volts and Amp-hours, and everything should be fine.

If you connect Solar Panels in series, a PWM Charge Controller is needed. More on that in the specific chapter.

In a parallel setup, the system will still be able to produce even if one panel is defective.

However, even ten 17 V Solar Panels connected in Parallel will not be able to charge a 24 V battery set made of two 12 V batteries wired in series.

As we said before, to get the most out of your system, the best choice is always to join together panels with the same characteristics (Voltage and Amperage) and possibly from the same manufacturer. If you wire Solar Panels with different specs in Parallel, Amps will add-up, but the Voltage output will align to the lowest panel.

Series + Parallel

A combined system, or array, with both series and the parallel connection, is the go-to solution when building a higher Voltage system with at least three Solar Panels.

This kind of setup will allow you to increase both the Current supply and the Voltage (in case you need it).

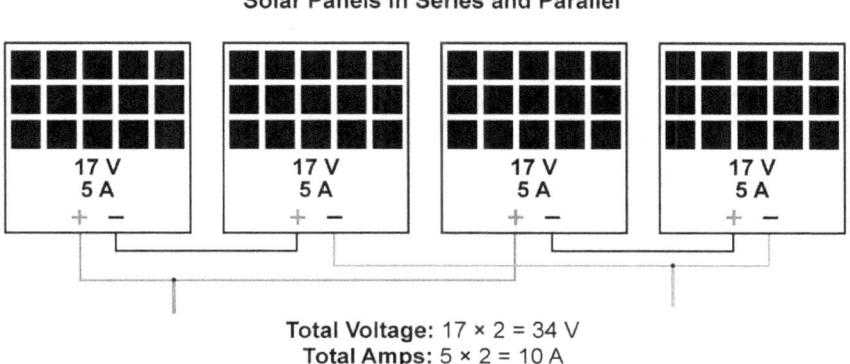

For the subject covered in this book, that is a standard off-grid PV system, it is unlikely that you'll need more than a 12 V Voltage, at least in the beginning. Therefore, we will not investigate series+parallel arrays further than this.

Combiner Box

Whatever wiring type suits your needs best, you'll find yourself with lots of wires everywhere since each Solar Panel has a red (positive) and a black (negative) wire.

To avoid ending up tangled in a jungle of vines, you'll want to use a combiner box. A combiner box is a little box that enables you to reduce the many wires coming from your panels in just two comfortable wires, so it is easier for you to build the rest of your system.

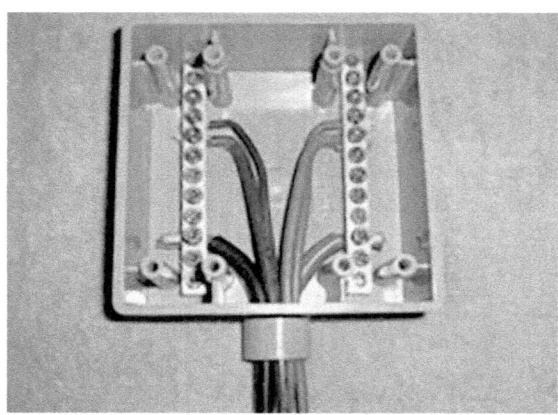

Choosing Your Solar Panels

Solar Panels come in different sizes, shapes, technologies. This can get a bit confusing, especially for absolute PV beginners. Let's break this down together!

Cell type

Today there are mainly three types of technology when it comes to photovoltaic panels: Monocrystalline, Polycrystalline, and Thin-film. Each one has its own advantages and disadvantages, so the final choice will depend on your unique needs entirely.

Monocrystalline

Monocrystalline solar cells are made of silicon wafers. These wafers are assembled into tiles and covered with a glass sheet. Monocrystalline differentiate from Polycrystalline cells as they rely on one pure crystal of silicon. This type of panel usually has black cells, not blue, because the light interacts differently with pure silicon crystal.

Monocrystalline panels typically have the highest light-to-electricity efficiencies of all PV panels (20% on average). They also have the greatest power capacity. Not only Monocrystalline panels are able to produce more electricity than Polycrystalline and Thin-Film, but they also usually come with higher wattage (300-400 W).

This is why Monocrystalline technology is perfectly suited for mobile applications such as RVs, boats, and cabins. With more electricity per cell, these panels allow optimizing the reduced space available.

The only downside is: they have higher costs.

Polycrystalline

Polycrystalline solar cells are also made of silicon wafers assembled into tiles and covered with a glass sheet. They differentiate from Monocrystalline cells as they are made of fragments of silicon crystals instead of being one single pure crystal. They are the classic blue cell panels.

Polycrystalline panels have lower light-to-electricity efficiencies of all PV panels, around 15% on average, but they come with more "human" prices.

This is why Polycrystalline modules are often preferred over Monocrystalline for mobile applications.

Thin-Film

Thin-Film panels are, as the name suggests, very thin, lightweight, and flexible. To achieve these characteristics, they cannot rely on silicon crystals but can be made of a variety of different materials, usually placed between two layers of transparent conductive material which job is to gather the light.

Thin-Film panels can be made of amorphous silicon (a-Si), cadmium telluride (CdTe), copper gallium indium diselenide (CIGS), or organic solar cells (OPV). Therefore efficiency would differ from case to case.

Amorphous cells have an average efficiency of 7%, while Cadmium telluride and OPV both reach a good 11%. The best technology for Thin-Film is CIGS, which an efficiency very similar to

Monocrystalline cells: around 20%. Problem is, CIGS is extremely expensive, even more than Monocrystalline.

If you are considering Cadmium Telluride technology, which is one of the cheapest out there, be aware that cadmium is highly polluting and poses a number of recycling issues.

OPV can come in different colors and can even be transparent, but its lifespan is shorter due to organic degradation.

Solar Tiles Anyone?

Solar tiles for your roof can be a solution if you need to power-up a cabin. Some can reach a good efficiency of around 17%, but they are very, very expensive for a second-house solution. And you usually can't install them on your own.

Rigid or Flexible?

When designing your PV system, you shouldn't be focusing solely on panels' efficiency but also paying attention to the setup.

In the next pages, we'll do just that.

Solar Panels can be rigid or flexible, so which one will you choose? Well, it depends. Let's take a look at the options and which use they are most suited for.

Rigid Panels

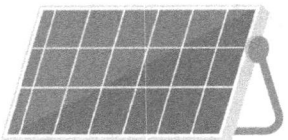

Rigid panels are modules mounted on an aluminum frame and covered with tempered glass. They are designed to withstand any weather conditions, as they will be used extensively outdoors.

The tempered glass is resistant to scratches, which will help maintain the panel's efficiency and clean them when they get frosted.

This durability allows manufacturers to give a long warranty, usually ranging from 10 to 30 years.

Rigid panels come in any size you can imagine, and they are easier to aim to the sun.

Flexible Panels

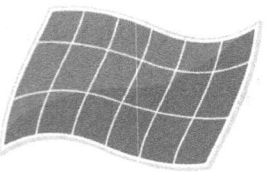

Flexible panels are modules arranged in a flat sheet of cells. They don't have a hard frame around them. Therefore they are much lighter than rigid panels, and they are designed to bend if needed.

They are covered with a plastic film to protect the solar cells, but this is not as protective as tempered glass, so the surface can get some scratches that will reduce efficiency. On the other hand, flexible panels are far less prone to crack in case of bigger impacts.

These panels are usually bendable up to 30 degrees, which is very useful if you have to fit them on a curved surface as boats and vans often have.

However, they are far less durable, and so manufacturers usually provide a shorter warranty.

Unless you don't mount them on a permanent stand, sun exposure can be a problem, as they are not free-standing. Also, when bent, a smaller surface will get direct light, and therefore production will be less efficient.

Fixed or Portable?

The choice between fixed and portable Solar Panels is entirely dependent on what usage you need them for. Want to mount them on that cabin in the woods you visit once a year? Maybe it's not that wise to let your panels out all year long…

But let's look at some generic pros and cons.

Fixed Panels

Fixed panels are permanently mounted on your boat, van, camper, cabin, etc.

This makes them harder to steal and enable you to produce electricity all-day-long whether you are there or whether you're not! One possible drawback is that they can't be seasonally adjusted to the sun angle unless you use a Tilt Mount (more on that later).

For RVs, a major aspect to consider is that if you want energy to be produced, especially if you haven't set up a large enough battery bank, you'll have to park in the sun, which may not be that pleasant in summer. Also, PV panels mounted on top of the van will raise your roofline.

Portable Panels

Portable panels are not permanently mounted but pulled out when needed. This may be a good idea if you are afraid someone could come and steal your panels while you're out. On the other hand, when you pull them out, you are then bound to keep an eye on them, as portable panels can be very easily stolen. Especially lightweight Thin-Film modules.

You'll also have to allocate a special space in your van, camper, or boat to store them, which can be tricky. Especially since you are not getting a constant flow of energy, and you may need more and bigger panels to compensate.

If you are planning a trip, consider that you'll have to manually mount and dismount them several times a day, and you'll need to be very careful if you do so on the roof to harvest sunlight while driving.

On the other hand, this allows for more flexibility as you could design a customized set up with multiple options to get your panels to always face the sun. Another interesting aspect is that you can park in the shade while your panels are collecting energy in the sun.

Weight

Weight is another key aspect to consider when designing your solar power system. Be sure to check how much weight your roof or boat can sustain, and then choose your panel, or panels, accordingly.

On RVs, excessive weight, even if supported, can increase fuel costs.

Size

Solar Panels come in every size, so you won't be low on options. Be sure to calculate the actual space you have at your disposal before buying your panels.

As we covered before, the available space is most likely to dictate the type of panel you'll have to buy to effectively fulfill your energy needs. Less space will automatically push you towards more efficient panels.

Wattage

Talking about efficiency, you'll also have to figure out how much Power a solar panel will produce. As we covered before, Power is expressed in Watts, so when you hear the word "wattage", it basically means Power.

In our case, we are not looking at Power consumption (a 60W lightbulb needs 60W to work), but rather at Power production. Therefore the higher the wattage, the best, since the cells and panel will be more productive.

Remember to check the wattage rating for both angled and flat installations.

Get the Most Out of Your Panels

Temperature

It is a common mistake to think that southern countries are luckier when it comes to solar. Well, not so fast.

There's not just sun and irradiance. Temperature, too, affects the solar panel power output. And we are talking of both external and internal, or cell, temperature.

The more Current flows into a cell, the hotter it gets. We already discussed these matters in the battery and circuit chapters, and the same principles apply here. Some energy converts into heat, thus causing power loss. Anyway, there's not much you can do. As the panel produces energy, the cell will inevitably heat.

As per external temperature, solar panels work best on cold sunny days. Hotter temperatures drastically decrease a panel's efficiency. That's why southern countries are not necessarily luckier with solar.

Here's a graph that shows just how much efficiency is lost as the temperature rises.

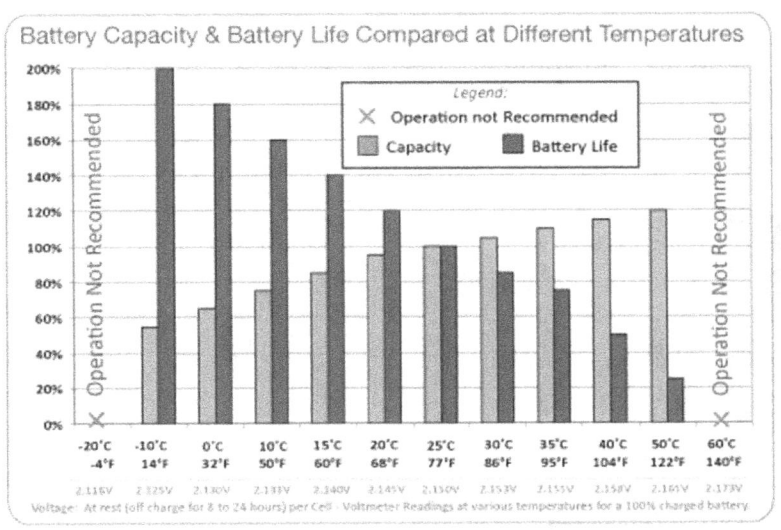

How to minimize efficiency loss, though?

Well, you'll have to consider two things. One is the temperature coefficient. The other is allowing air to circulate above and below the panel to cool it down by mounting it a few inches off the roof.

The temperature coefficient is a simple parameter that you'll find by looking at the datasheet of a solar panel, often referred to as Pmax. Its value is given in the form of a negative percentage that illustrates how much efficiency is gained or lost for each 1-degree change in temperature. When the temperature goes up, efficiency goes down. When the temperature goes down, efficiency goes up. The starting point for calculating the temperature coefficient is 77°F (25°C), which is the temperature solar panels are tested at. This is also known as STC, or Standard Test Conditions. Most solar panels have a temperature coefficient ranging from -0.3%/°C to -0.5%/°C.

Let's make an example with a Solar Panel that reads:

- Efficiency: 14.8% STC
- Wattage: 260 W
- Temperature coefficient: -0.5%/°C
- External temperature: 100° F

First we need to convert Celsius degrees into Fahrenheit.

77°F = 25°C

100°F = 37.7°C

Now let's make our calculations:

37.7°C − 25°C = 12.7°C

(Increase in temperature)

12.7°C × −0.5% = −0.0635 or −6.35%

(Loss in power output at 100°F)

If you apply this value to the Solar Panel's Wattage you'll get the effective power loss at 37.7°C (100°F).

260 W × −6.35% = 16.51 W

(Power loss at 100°F)

260 W − 16.51 W = 243.49 W

(Actual Watts produced at 100°F)

Note that some manufacturers will provide a temperature coefficient based on NOCT testing instead of STC. In that case, the base temperature for our calculation will be 68°F (20°C).

There are a number of other little specs, but I'm trying to keep it as simple as possible, and it's already a lot of information to digest. These are the essential things to know, though. As long as you keep them in mind, everything will be fine.

Also, remember to install the panels at least a few inches off the roof to allow airflow, and thus enable the PV module to cool down and preserve efficiency more effectively.

Shading

Shading loss is maybe the most common issue when it comes to solar power production.

Your panels can get shaded by trees, mountains, other building's shadow, and many other things that can vary from hour to hour in a day. This is called far-shading, and it will negatively affect your production. Most of them are inevitable far-shading. Nearby objects, on the other hand, are much more annoying since they can even stop your production completely.

Did you know that even a small shadow in the corner of one panel can ruin the entire production of a panel? This completely struck me when I was first starting to learn about photovoltaic.

In fact, this is pure logic if you think about it.

Shading affects irradiance, which affects the Current output of the module. If your PV modules are connected in series, you already know that the overall Current (or amperage) will align to the lowest panel. Therefore, even a small shade on a single panel will reduce the string's Current.

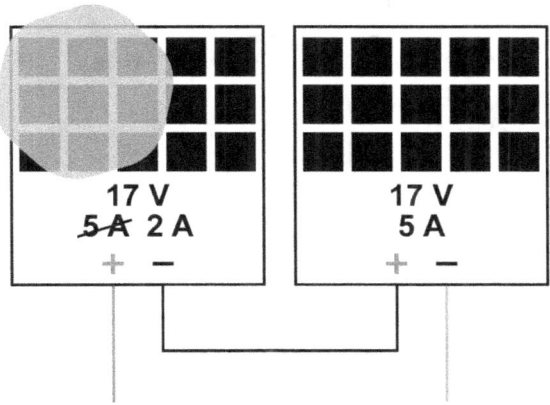

Fortunately, most manufacturers install 3 or more bypass diodes in the backside of PV modules to minimize this problem. Another trick is to carefully calculate fixed shadow projections, like the one from a sail on a boat, for example, and place the panels in a way that it gets the less shade possible.

In Parallel wiring, Solar Panels are spared from this kind of issue, but shading can cause back-feeding, which means basically that electrical power flows the wrong way (never a good idea).

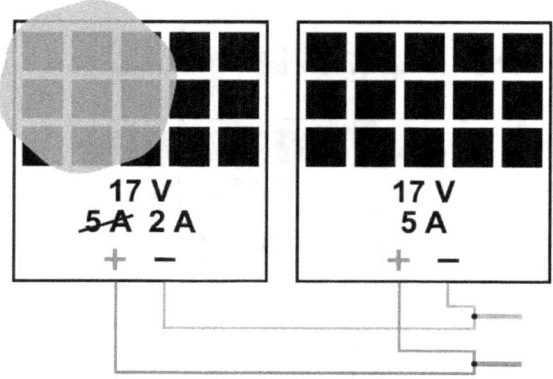

Total Voltage: 17 V
Total Amps: 7 A

Again this can be prevented by installing blocking diodes, which again are often already built-in from manufacturers.

In either case, be sure to check the datasheet or ask the manufacturer.

Azimuth and Tilt Angle

The Azimuth angle is the direction from which the sunlight comes. If we are talking solar, then the Azimuth angle is meant to point out the direction of Solar Panels regarding the sun's orientation.

The reference here is solar noon when the sun is directly North in the southern hemisphere and directly South in the northern hemisphere. That's why in order to harness maximum power, you Solar Panels should face South in the northern hemisphere or face North in the southern hemisphere.

Of course, this only makes sense if you are mounting PV modules on a cabin or house roof. RVs and boats being continuously moved, determining the Azimuth angle is entirely useless. Same goes for flat panels.

If you're using portable panels, though, you can use a compass to determine the Azimuth angle and place the modules facing South or North based on the hemisphere you find yourself into.

There are a lot of formulas if you want to calculate the Azimuth angle, but I'm not going to cover them here since it would only add useless jargon to the topic. I am not here to compile the "Solar Bible," but rather a quick and easy manual to get started with solar power and get you up and going with your 12V PV system.

Anyway, you can always rely on this useful Azimuth calculator from Casio: *https://keisan.casio.com/exec/system/1224682277*

Apart from the Azimuth angle, there's the Tilt angle, which is related to the latitude of your location.

Here the key is to have an idea of when the PV system will be used mostly since the altitude and direction of the sun vary with the seasons.

In off-grid applications, the Tilt angle will depend on what you need the panels for. If you are powering an Rv, then it is most likely that you will be using it in the summer, so it is better to adjust the angle for that season. The opposite is true if you are planning to use your camper or cabin during the winter. Finally, if you are going to live the van life, or for whatever reason, use your off-grid PV system all-year-long, it is preferable to optimize the angle for the winter season since it will be more difficult to gather enough power to cover your energy needs.

Again, there are a number of calculations involved, and I will not cover them here.

Anyway, you can find the optimal tilt for your situation seasonally or month-by-month using these online calculators:

- *https://www.foresthillweather.com/PHP/Conversion/SunCalc2.php*
- *https://www.shaktipumps.com/solar-calculator.php*
- *https://magesolar.com/solar-angle-calculator/*

To find out your latitude, you can open Google Maps from your computer. Right-click the place or area on the map and select "What's here?". A card will then appear at the bottom with the coordinates. The first one with the N is your latitude.

Safety

Solar Panels are required to be grounded/earthed to establish an effective ground-fault Current path.

Grounding/Earthing could save your life, avoid fires, and anyway, it is compulsory for every PV system for obvious safety reasons. Feel free to check the Basic Circuits chapter for a recap.

Earthing a Solar Panel means to connect one side of a copper wire (a green wire) to the metal structure of the panel by using a wire lug and then connect the other to the earth.

If you are mounting panels on a cabin or house, you'll need to wire the panels to a ground rod, which is a copper stick planted 8 feet deep into the ground. If the cabin is already connected to the grid, then you are required to connect the Earthing green wire to the existing Earthing system of the house, which is called bonding.

If you are mounting panels on RVs or boats, the green wire coming from the metal structure of your panel should be connected to the chassis of the vehicle.

If you have multiple panels, you can use one single wire to connect them all, but most of the time, each panel comes with its own green wire.

Should You Buy Used Panels?

Well, it depends.

Buying used panels could save you a lot of money, but you have to do it carefully in order to avoid taking in bad or damaged panels that will end up costing you more money in the long-term.

Like everything that is bought used - except batteries! Never purchase used batteries! -you can find some incredibly good deals and some terribly wrong deals.

If you are not an expert, or if you want to make things quick, I recommend buying new panels. They often outmatch older panels in performance and often come at a lower price than those old panels when they were new.

Anyway, if you want to experiment with used panels, keep in mind that you cannot tell by the eye if they're in good shape or not. Unless they have clear damage signs on some cells, which will obviously reduce the power output. Checking the Voltage with a voltmeter can be a good idea, but again, it's no guarantee. Also, consider that Solar Panels' efficiency drops from 0.5% to 1% every year, so don't buy panels that are too old. In general, do your research and check the actual datasheet of the used panel you are considering, and if you can get some help from someone with more expertise in the field.

In my experience, though, and with a little help from my dear friend Jim, I found out that some people are more interested in getting rid of their bulky panels than to get the right price out of them.

Charge Controller

Meet your system's caretaker: Mr. Charge Controller!

Solar Panels and Batteries deal with varying Voltage and Current. We've already covered a whole bunch of common issues they can encounter.

This is where Charge Controllers come into play.

Charge Controllers protect your PV system, and especially your batteries.

They handle 3 main tasks:

1. Monitor the State of Charge (SOC) of your batteries.
2. Regulate the Current flow to avoid over-charging and over-discharging.
3. Adjust the Voltage of the PV modules to the required Voltage of the batteries.

But Mr. Charge Controller is no lazy guy and offers other useful services as well.

Short-Circuit and Overload Protection

Modern Charge Controllers have built-in overload protection that will break the circuit to avoid damaging the system and prevent fire hazards.

Blocking Reverse Current

Solar Panels pump Current in one direction, but at night they stop pumping and instead draw a bit of Current from the Battery. Solar Chargers can prevent this kind of back-feeding.

Temperature Monitoring

As we now know, temperature heavily affects batteries. Many Charge Controller can compensate for temperature and adjust Voltage set points to maximize efficiency. They do so through a built-in sensor (in this case, the device should be positioned in a place with a temperature similar to the Battery spot) or via a sensor-cable directly connected to the Battery.

Display and Metering

Imagine having to drive your car around without a clue on fuel levels, speed, etc. It would be impossible. Same goes for a PV system. Information is essential to know what's going on and whether everything is working fine or not. Most Charge Controller today come with a display indicating SOC, flow, and other important indicators.

PWM and MPPT

There are two types of Charge Controllers. PWM, or Pulse Width Modulation, and MPPT, or Maximum Power Point Tracker.

PWM

PWM controllers act just like a switch between Solar Panels and Batteries. It adjusts the Voltage by opening and closing the "switch" multiple times in a row. This is why it's called Pulse Width Modulation.

PWM controllers are cheaper and less efficient since they don't operate at the maximum power point but rather set the Voltage close to the Battery's nominal Voltage.

MPPT

MPPT controllers are, in fact, a DC-DC converter. That is a device that transforms a DC signal into another DC signal but with different parameters.

As the name suggests, MPPT controllers have a tracker that measures the exact maximum power point of the Current / Voltage curve, thus enabling more efficient performances.

MPPT controllers are far more expensive than PWM, but they are more efficient (up to 30% more Power than PWM). They are essential if you need to wire multiple panels in series.

Selecting a Charge Controller

The Right Fit for Your System

As usual, first thing is to check the datasheet. You have to make sure that the Charge Controller you are considering is able to deal with your specific Battery and Solar Panel. And this will imply some calculations.

Model	ROV-20	ROV-40
Nominal system voltage	12V/24V Auto Recognition	
Rated Battery Current	20A	40A
Rated Load Current	20A	20A
Max. PV Input Short Current	25A	50A
Max. Battery Voltage	32V	
Max Solar Input Voltage	100 VDC	
Max. Solar Input Power	12V @ 260W	12V @ 520W
	24V @ 520W	24V @ 1040W
Self-Consumption	≤100mA @ 12V	
	≤58Ma @ 24V	
Charge circuit voltage drop	≤ 0.26V	
Discharge circuit voltage drop	≤ 0.15V	
Temp. Compensation	-3mV/°C/2V (default)	
Communication	RS232	
Battery Type	Sealed (AGM), Gel, Flooded and Lithium-ion	

First, Nominal Voltage. Make sure that the nominal Voltage matches the Battery. Typically it is rather 12 V, 24 V, or 48 V. Especially if you are building a larger Voltage Battery Bank, verify that the Charge Controller is actually able to deal with that amount of Voltage.

Second, Current (aka Amperage). It's the "Rated Battery Current" line on the Charge Controller's datasheet. To evaluate how much Current the Charge Controller will have to deal with, we are going to be using the Power formula.

```
P = V × I
V = P ÷ I
I = P ÷ V
```

Let's say you have a 400 W solar array and a 12 V Battery Bank.

```
400 W ÷ 12 V = 33.33 A
```

In reality, the actual Battery Voltage is a little higher than 12 V, usually around 12.7 - 12.9 V. So let's do this again.

```
400 W ÷ 12.9 V = 31 A
```

It is also preferable to use the safety factor of 1.2 to prevent any future trouble. Let's do this.

```
31 A × 1.2 = 37.2 A
```

Therefore, you'll need a controller capable of at least 37.2 Amps.

Most controllers are rated 40, 60, 80, or 96 Amps, so in this case, you would need to choose a 40 A Charge Controller. Generally speaking, always select the Charge Controller with the next higher rating than your actual Current need.

Consider that a higher Voltage Battery Bank will need less Current. For example, let's say we have a 24 V Battery Bank this time:

```
400 W ÷ 24 V = 16.6 A
```

Third, Power (aka Wattage). It's the "Max Solar Input Power" on the Charge Controller's datasheet. Well, the label is kind of self-explanatory here. You just have to make sure that your Solar Panel array's Wattage is manageable by the Charge Controller.

Fourth, Solar input Voltage. It's the "Max Solar Input Power." Here you'll have to consider the Voltage of your Solar array. Usually:

1 Panel = 17 V

2 Panels in parallel = 17 V

2 Panels in series = 17 × 2 = 34 V

And so on.

Just make sure that the Charge Controller can deal with the Voltage coming off your Solar array.

PWM or MPPT

MPPT Charge Controllers are the most advanced technology, and they are the go-to for any bigger or more complex PV system.

But here we are talking about a simple 12 V off-grid system. Therefore PWM can be considered.

Apart from the cost issue, there's just one fundamental reason to choose MPPT over a PWM. And that reason regards to wether or not you are planning to connect multiple Solar Panels in series.

If this is the case, then you'll need an MPPT. Otherwise, a PWM can suffice.

As you now know, PV modules wired in series add up their Voltage. Therefore the Solar Charger will need to convert the greater Solar Panel Voltage to match the lower Battery Voltage. The more the distance between the two Voltages, the more the power losses. That's why you'll need a Maximum Power Point Tracking device to ensure adequate efficiency. Of course, the distance can be reduced with a higher Voltage Battery Bank.

Temperature compensation

Some Charge Controllers have fixed temperature compensation, while others have an adjustable setpoint. If you are not experienced, I suggest not worrying too much about this topic. Charge Controllers today are pretty good and will do a good job even if you are not manually adjusting temperature compensation setpoints.

Display (external display)

Having all the information about Current and the Battery's SOC is crucial to run an off-grid PV system. Therefore I recommend you buy a Charge Controller with an incorporated display.

If, for whatever reason, you end up with a Charge Controller without a display, external devices can be easily bought online. In fact, many of them provide more information and metrics than standard displays, so it can be a good idea, while not necessary, to buy one anyway.

Low Voltage Disconnect

A Low Voltage Disconnect (LVD) is a safety feature that prevents over-discharging. While all Charge Controller protects your Battery from overcharging, not every Charge Controller can handle over-discharging issues.

Make sure your device has a built-in LVD and that this specific LVD can handle your DC loads (if you have any connection to your Charge Controller, which I don't recommend).

In any case, you can also buy a separate LVD device or install blocking diodes, as I mentioned in the Solar Panel's chapter.

Inverter

As you may have noticed, both Batteries and Charge Controllers operate with DC outputs. So how will you be able to power all your AC appliances (typically refrigerators, dishwashers, toasters, microwave ovens, ...)?

Let me introduce you to another friend of ours: Mr. Inverter.

He is a very private guy. He'll be more than happy to convert your DC signal into that AC signal you need to power the loads that generally operate on power supplied by the utility grid.

Best Inverters for Off-Grid Applications

There are different types of Inverter out there for the most diverse application. The most common type is meant to work on grid-tied applications. Therefore, we'll need to narrow down the kind of Inverter we want to consider for our off-grid application.

Off-Grid Battery-Based Inverter

Off-grid battery-based Inverters are specifically designed to convert DC electricity from a battery bank to an AC signal.

This kind of Inverter will have an input value matching the usual nominal Voltage of the Battery, which is 12 V, 24 V, or 48 V.

In most cases, a battery-based Inverter will also provide some kind of overload protection.

Marine Inverter

Marine Inverters are similar to battery-based Inverters since they both deal with off-grid applications. The only difference is that they are designed to withstand a typical marine environment, such as salt corrosion.

If you are building an off-grid system for a leisure boat, then marine Inverters are the go-to Inverter for you.

Inverter-Charger

Inverter-Chargers have an additional feature that can be very useful. In addition to working one-way, that is, taking the DC power from the Battery and converting it to an AC signal, it is designed to work both ways.

In fact, with an Inverter-Charger, you'll have an additional "line" to charge your Battery without relying solely on Solar Panels. You'll be able to use shore power or plug a generator to the Inverter-

Charger, turn the AC signal into DC, and get that power to charge the Battery bank.

Some Inverter-Chargers even offer the possibility to start a generator remotely when the Battery's charge level gets too low.

Output Signal Types

Another aspect to consider is the output signal of the Inverter.

Square Wave and Modified Sine Wave

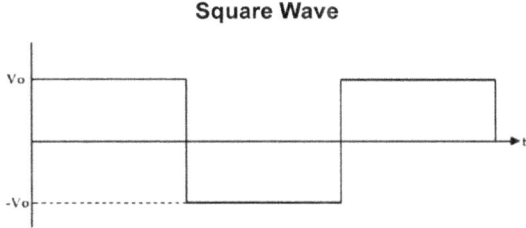

Square wave Inverters are the oldest technology available and the most affordable Inverters on the market. On the other hand, they make a terrible noise and heat up most loads so much that they cannot be considered for applications such as laptop computers, appliances with microprocessor controls, laser printers, and so on. In many European countries, square wave Inverters are banned today.

So we are going to discard this type of Inverter right off the bat and look for other solutions.

Modified sine wave Inverters are, basically, enhanced square wave. They add up some levels to simulate a sine wave.

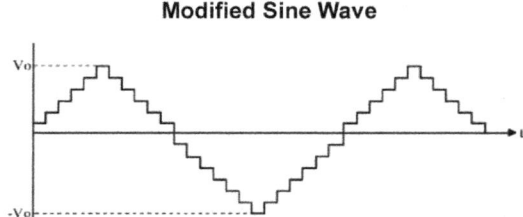

Modified Sine Wave

They are more expensive than square wave ones because they can support a larger number of appliances, but they also inherit many of the drawbacks from the square wave type. Especially MSW with lower square levels makes a lot of noise and can heat up your loads.

MSW Inverters are not recommended to power laptop computers and battery chargers. They even make some background buzz with most audio equipment. Therefore I guess they are not that useful in the 21st century, right?

In Europe, MSW needs to be paired with a PFC or Power Factor Corrector to reduce harmonics.

We're going technical here, and that is not the purpose of this book.

Let's make it short: what you need, in order to safely operate whatever you need, is a pure sine wave Inverter.

Pure Sine Wave Inverter

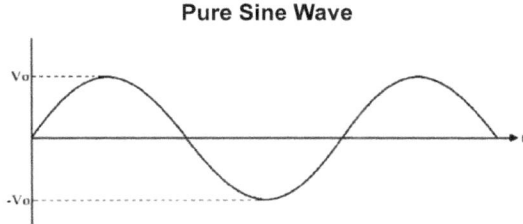

And here we are. Pure sine wave Inverters are the only possible choice if you want a stress-free PV system.

A perfect sine wave Inverter provides the exact same signal as the utility grid that powers our houses, which means everything will work fine.

Of course, higher technology and higher efficiency will translate into higher prices. Still, PSW Inverters will not heat up or burn your chargers and appliances, and they will not make that much noise.

They are safer, ensure you can plug any appliance you'll possibly need, they'll draw less power to them, and ensure a longer lifespan for your loads.

Selecting an Inverter

Things to consider when choosing an Inverter.

Efficiency

Today, most Inverters have efficiency values over 93%, which is good. But you can find Inverters with even higher ratings, going close to 100%. I suggest you set your expectations around 95-97% to have fewer energy losses at a competitive price.

Output Voltage

Most Inverters supply 120 V (AC), but more and more Inverters are now able to provide 120/240 V power.

It is always better to check with the manufacturer. Still, many Inverters today can even be wired in series for increased Voltage output or in parallel for increased Current (A) output.

Peak Power

Like any other electrical component, Inverters have limits, too. Always be sure to check how much power the Inverter is able to provide during a short amount of time.

Here you need to consider different time frames. For intensive use, we usually consider the amount of power that you can draw over 30 minutes (usually 2,500 W). For sudden peaks, you'll need to check how much surge power the Inverter is able to sustain for 5 seconds.

Position

When considering an Inverter, the size can be a critical factor as well, since you'll need to place it near the Batteries.

Wires and Stuff

Well, it's very nice we have all our components lined up here. We've got the Solar Panels, we've got the Battery, the Charge Controller, and the Inverter. But if we are ever going to use them, we miss the highways to connect them all together in a beautiful circuit: the wires.

Here's how we are going to choose them and use them.

The Right Wire Type

Copper or Aluminium?

You can either use copper wire, ore aluminum wire. Copper wire is a little more expensive, but it has fewer power losses. It is more flexible than aluminum, and it is better at preventing overheating.

Solid or Stranded?

Basically, solid wires contain a single metal wire core, and stranded wires contain multiple stranded wire cores.

Stranded wires usually have a larger diameter, better flexibility, and better conductivity but come at a higher sales price.

I recommend stranded wires (at least for outdoor cables), but this is not compulsory for your PV system.

Insulation

Insulation is another critical factor when it comes to wiring. Your wires will be exposed to a variety of weather conditions, especially if you are building a PV system for boats and RVs. Therefore you should consider UV, water, chemical, and heat resistance.

Wires' names indicate what they are built for. Looking at those acronyms like that could be a little disorienting, but when you know their actual meaning, they will help you out.

- H stands for Heat resistant. Sometimes found as HH, or High Heat.
- W stands for Water-resistant.
- R stands for Rubber insulation.
- T stands for Thermoplastic insulation.
- X stands for XLPE, that is Cross-Linked Polyethylene.
- N stands for Nylon insulation.

Now here are the different wires we can choose from:

- THHN, ideal for dry conditions. Withstands 194°F (90°C), not water.
- THWN / THWN-2, ideal for conduit applications.
- RHW / RHW-2, ideal in outdoor applications.
- XHHW / XHHW-2, have better resistance to chemical exposure and abrasion.
- USE-2 (Underground Service Entrance), ideal for underground wet conditions, can withstand higher pressure.

And then there is:

- PV Wire is designed explicitly for wiring Solar Panels since they can withstand water, extreme UV and heat exposure, and an extra layer of insulation.

Color

Despite the color difference, all wires are equal (if they are of the same type). But we use a specific color code to make maintenance more comfortable and safer.

DC Power

- Red: Positive Current wire.
- Black: Negative Current wire.
- Green: Earth (sometimes green with yellow stripes)

- White/Grey: Ground

AC Power *(120/208/240 Volts)*

- Black: Phase 1
- Red: Phase 2
- Blue: Phase 3
- White: Neutral
- Green: Ground/Earth (sometimes green with yellow stripes)

WARNING: This is true for the USA National Electrical Code (NEC). Always be sure to check the right color code of your country.

The Right Wire Size

Choosing the appropriate size of wires and cables is critical in every PV system. The right wire reduces energy loss and prevents overheating. And in fact, using undersized cables violates the National Electric Code (NEC) and will most probably bring you troubles

Thickness and Rating

First off, we are going to talk about the thickness, which is closely related to a wire's rating.

The rating is what you will want to check since it's what determines how much Current the wire can handle safely. In the US, it is determined on the American Wire Gauge (AWG).

Basically, the AWG indicates (on its own scale) the resistance of the wire. The lower the AWG, the less the resistance. And as you now know, the lower the resistance, the higher the Current (Amps) it can deliver.

To deliver more and more Amps, a wire needs to have more and more metal wire cores. That's why a lower AWG rating means a thicker wire.

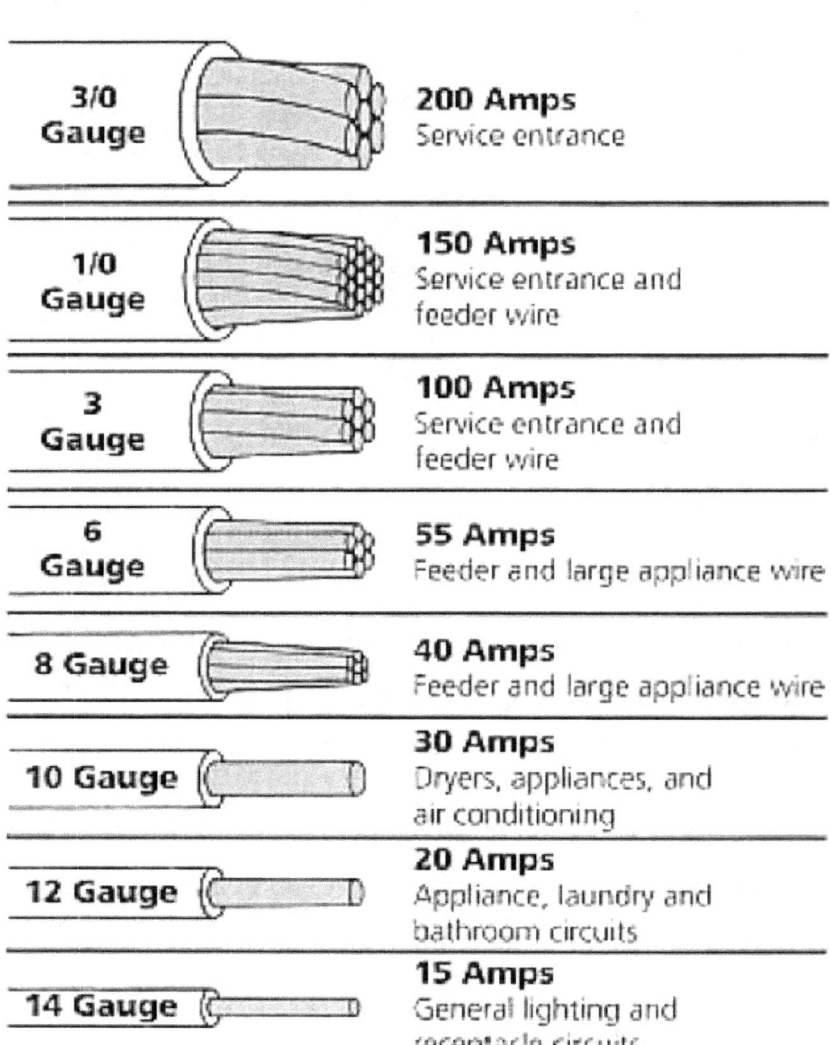

Length and Voltage Drop

You would think a wire is a wire, no matter how long it is, but this is not actually the case.

In fact, the longer the wire, the less the efficiency. The longer the distance electricity travels, the higher Current is expended. Which means Voltage drop.

With some calculation, we will be able to keep that Voltage drop under 2-3%. But we will cover that later, in the Doing section.

The Right Wire for the Right Use

Solar Panels to Combiner box

Since it will be an outdoor application, the best choice here is a PV Wire, but USE-2 and RHW-2 are valid alternatives.

In the Doing section, we will be calculating everything upon your necessities, but as a general rule-of-thumb, each PV module can be paired with a 10-12 AWG wire.

When you buy Solar Panels, though, they often come with their own cables, which are usually designed for this specific use. No need to go looking for other wires.

Combiner Box to Charge Controller

Here you can stop worrying about water. From the Combiner Box on, your wires will most likely be running in dry conditions. Therefore, THHN-2 will suffice.

Of course, if you are building a PV system for your boat, water will always be a security issue, so you may want to consider THWN-2 as well.

Here you will be looking for thicker wires with a lower AWG rating (3-8) since they will need to carry the Current coming from multiple Solar Panels.

Charge Controller to Batteries

Since the Battery Bank spot can get very hot, you will need wires that can withstand higher temperatures. Therefore, I recommend you use a HH wire type, like THHN-2.

Moreover, Batteries are made of chemicals, so you can also consider XHHW-2 wires for extra safety. When they fail, lead-acid batteries also produce water, so a water-resistant wire is always a good idea, too.

On boats and in wet conditions, again, be sure to choose a water-resistant wire.

As per the Current needs, keep the same gauge and diameter of the wires coming from the Combiner Box.

Batteries to Batteries

We're still talking about Batteries, so the same recommendations are valid here. Use a wire type with high heat, water, and chemical resistance.

As the Batteries will be connected to the Inverter, which is connected to the loads. And as we can safely presume that higher Current demand will happen sooner or later. I'll suggest you use even thicker cables, ranging from 3 to 3/0 AWG.

When connecting multiple Batteries together, be sure to use wires of the same length, same diameter/AWG, and from the same manufacturer.

Batteries to Inverter

Same as above. We are still in the Battery Bank's case.

Inverter to Loads

Apart from boats and very wet environments, we can safely state that there are dry conditions inside the living space of your camper truck or cabin. Therefore THHN-2 cables are more than enough.

Here a smaller diameter wire suffices, too. You can safely choose a 12-14 AWG gauge.

Fuses

Fuses are essential to protect your circuit from overheating, melting, and even catching fire.

Not all fuses are equal, though. Fuses sizes have to be based on the wire size you'll be connecting them to.

Fuse size for Solar Panels' wires is often indicated on the datasheet.

Remember to place the fuses on the red (positive) wire and as close as possible to the energy source.

Busbars

Busbars are a piece of technology that makes complicated power distribution easier and more flexible. In fact, they distribute electricity just like switchboards distributed phone calls.

They are often made of copper or aluminum and can ground electricity.

In PV systems, they are often used to distribute DC Power to the DC appliances.

Shunts

A shunt is a current resistor or ammeter. Basically, it is a high precision resistor that enables us to measure Voltage and Current by dividing the input Voltage by the shunt's own resistance (thank you, Mr. Ohm). To do so, shunts have to be connected in series on the DC line.

Voltage, Current, Resistance. That's a bunch of useful data, right?

Precisely! And that is why, in PV systems, you often use shunts with Battery Monitors in order to get accurate readings of Current, Voltage, State of Charge, and stuff.

How to choose a shunt?

First, the maximum Current that the shunt can bear has to match the Current coming out from your Battery Bank.

Second, if you are using a shunt to pair it with a Battery Monitor, as it most likely is, the shunt has to match the Battery Monitor's specs, too.

Note: don't use a shunt with a higher rating than the load since the measurement will be inaccurate.

Circuit Breakers and Isolator Switches

While fuses sacrifice themselves to save the day, Circuit Breakers can automatically disconnect the circuit in case of surges and have a much longer lifespan. That's basically a panel with a series of switches, usually connected to DC loads. Every home is equipped with one, so you should be familiar with it.

Isolator Switches, on the other hand, are just, well, switches that you can manually operate to break a circuit. They are usually used to isolate components and allow maintenance.

National Electrical Code (NEC)

The National Electrical Code is the most credited American reference for electrical installations.

The NEC is updated every year, and you can get free access to it via the National Fire Protection Association: *https://www.nfpa.org/NEC/About-the-NEC/Free-online-access-to-the-NEC-and-other-electrical-standards*.

You can also look at the complete 2017 edition here: *https://www.tooltexas.org/wp-content/uploads/2018/08/2017-NEC-Code-2.pdf*.

There are specific chapters on recreational vehicles, and every kind of application you may have in mind. Thus I suggest you take a look.

Section 2
DOING

Designing the System

So far, we've learned a lot about how PV systems work, how to choose the right components, and how to get the most out of them.

Now it's time to put all this knowledge into practice! That's my favorite part!

A little disclaimer here. Each and every system and each and every system owner have their own specific needs. Covering every possible configuration would take, I guess, at least 1 billion books, and I don't know about you, but that sounds a little overwhelming to me.

I think we will be better off using a generic example on which to make our calculations. Let's do this!

For our example, let's say we want to build a 12 V PV system to power up an RV.

Sizing Your Needs

We are used to drawing infinite power from the grid. But when you rely on an off-grid system like the one we are designing here, the amount of energy we can draw from the sun and store in our batteries is limited.

That's why we are going to carefully evaluate what appliances we actually need, how much Power they need, and how long we are planning to use them.

For our RV example, I think we are going to be on the road or go on a hike pretty often. So I'm not going to take my everyday energy-consumption at home as a standard but rather size everything down to the essential. This is my personal rules-of-thumb, though. You can decide to design your system to get all the energy you want. Just keep in mind that the sun is king and out of our control and that the space on an RV's roof is limited, so you'll have to deal with these limitations.

Now make sure you've got a pen and paper (or an excel sheet opened), and let's get the ball rolling.

Appliances' Consumption Table

What you want to do here is to draw a table where to list the appliances you'll be using, their Voltage, Current, and Power consumption, and the estimated time of use per day. For clarity, I'll suggest you make separate lists for AC and DC loads.

Usually, you can find all these pieces of information on the devices themselves or on their chargers. You can also check the datasheet, check the manufacturer's website, or google the device.

Sometimes you've only got the Voltage and Amps information. Use our mighty formula to get the Watts:

```
P = V × I
```

Or translated into units:

```
Watts = Volts × Amps

Volts = Watts ÷ Amps

Amps = Watts ÷ Volts
```

I'll keep it simple here, and list just a few.

AC loads	Wattage	Hours / day	Watt-hours per day
Laptop charger (2)	85 W	2	× 2 units = 340 Wh/d
Phone charger (2)	5 W	2	× 2 units = 20 Wh/d
Hair dryer	1000 W	0.25	× 1 units = 250 Wh/d
			TOTAL (AC) = 610 Wh/d
DC loads			
Water pump	40 W	1	× 1 units = 40 Wh/d
Fridge	45 W	4	× 1 units = 180 Wh/d
Bath fan	25 W	1	× 1 units = 25 Wh/d
Lights (6)	19.2 W	5	× 6 units = 576 Wh/d
			TOTAL (DC) = 821 Wh/d
GRAND TOTAL			**(AC+DC) = 1431 Wh/d**

Note that to obtain the Watt-hours rate I made:

```
Watt hours = Watts × Hours × Load units
```

Also, note that the hours of use per appliance per day really depend on your habits and how you plan to live in your RV.

Calculating Battery storage

As I mentioned in the *Battery* section, you cannot discharge your battery beyond a certain rate. Never ever.

Lead-acid Batteries cannot be discharged beyond 50% of their Capacity.

Lithium Batteries cannot be discharged beyond 20% of their Capacity.

For these reasons, and since Battery Capacity is expressed in Amp-hours (Ah), we'll need some calculation here.

First, you'll need to decide the Voltage you want to use. In our example scenario, this is 12 V.

Then let's bring on the Amp-hours formula:

```
Amp hours = Amps × hours
```

That's not very helpful, right? Lucky enough, we know that Amps can be obtained with this other formula we used before:

```
Amps = Watts ÷ Volts
```

So in our scenario, that makes:

```
Amps = 1431 W ÷ 12 V = 119.25 Ah
```

Since our W value is actually Watt-hours, or Wh, our results will not be expressed in Amps but in Amp-hours, or Ah.

So now we know that we need 119.25 Ah to fulfill our energy needs. Cool.

As we know, Batteries can only serve a given percentage of their actual Ah rating. So let's calculate how much Ah our Battery Bank should provide.

Lead-acid Battery example:

Since Lead-acid Batteries can only be discharged at 50%, their actual Capacity is 50%. The ratio of the 100% Capacity, to the actual Capacity (50%) will result in our multiplier to find out how much Amp-hours our Battery Bank should provide.

```
100 ÷ 50 = 2
119.25 Ah × 2 = 238.5 Ah
```

Therefore, a lead-acid Battery Bank should be rated at least 238.5 Ah to provide the amount of electricity we need.

In this case, I will recommend buying 3 Batteries of 100 Ah each and wire them in parallel.

Lithium Battery example:

Since Lithium Batteries can only be discharged at 20%, their actual Capacity is 80%. The ratio of the 100% Capacity to the actual Capacity (80%) will result in our multiplier to find out how much Amp-hours our Battery Bank should provide.

```
100 ÷ 80 = 1.25
119.25 Ah × 1.25 = 149.07 Ah
```

Therefore, a lithium Battery Bank should be rated at least 149.07 Ah to provide the amount of electricity we need.

In this case, I will recommend buying 2 Batteries of 100 Ah each and wire them in parallel.

Real Life Use and Time of Discharge

Calculating the Amp-hours per day we need is not accurate enough, though.

The Ah Capacity of a Battery is almost always calculated on a constant usage over 20 hours at room temperature. But in real life, we'll probably use most of this Capacity in the evening, in a much shorter time span, with much higher intensity, and at a time where our Solar Panels cannot "refill" the Battery Bank.

What I mean by that is: it's much wiser to size your Battery Bank on the expected peak daily consumption rather than on the average daily Power use.

Let's get back to our example.

We can expect to have our Solar Panels constantly recharging our Batteries until 6PM, then the sun will set, and we will have to rely solely on the Battery.

We can assume our Battery Bank will be completely topped-up, and we can also expect to use the most electricity between 6 PM and 11 PM. That's 5 hours.

We've got 2 phones, and 2 laptops, that each needs one evening charge cycle of approximately 2 hours of charge.

Basically, in the peak hour, we'll probably be using each and every appliance listed above. We'll just have to cut down the laptop and smartphone charging demand since we're going to charge one device at a time. We only need one charge cycle, and we are only considering one hour.

That makes 565.2 Wh in the peak hour.

Let's find out how many Amp-hours that makes:

```
Amps = 565.2 W ÷ 12 V = 47.1 Ah
```

It looks that our 200 Ah Battery Bank (lithium) or 300 Ah Battery Bank (lead-acid) is more than enough, right? Wrong.

The Amp-hours value displayed on the datasheet is meant for a one-hour only usage to full discharge. Which is not something we want to do.

As we already calculated, a 300 Ah lead-acid Battery can serve us only 150 Ah (50% of Capacity), which is enough to cover our 47.1 Ah peak hour need.

But we have to consider that, at night, Solar Panels won't produce any Power to refill the Battery bank. Therefore, after the peak hour, you'll be left with 102.9 Amps for the remaining 4 hours, which means 25.73 Amps per hour.

Let's calculate how much Amp-hours we need to light our RV, charge one phone, and one laptop for one hour:

Lights:

```
19.2 W × 6 lightbulbs = 115.2 Wh

115.2 W ÷ 12 V = 9.6 Ah
```

Laptop + Phone:

```
85 W + 5 W = 90 Wh

90 W ÷ 12 V = 7.5 Ah
```

Total:

```
9.6 Ah + 7.5 Ah = 17.1 Ah
```

Sounds good so far!

Well, not so fast.

At this rate, we'll be emptying our Battery bank (to the safe depth of discharge of 50%) in 5 hours. Not 20 hours (or 10, given the max 50% depth of discharge).

That's where we want to check the C-rate of our Batteries.

In our case, we'll be considering a standard C_{20} rate (but be sure to check the specific C-rating of your Battery.)

A C_{20} rate means that the Battery will take 20 hours to discharge at 100%. But we only want to use half of our lead-acid Battery (or 80% of our lithium Battery), so first, let's cut the discharge time out.

```
20 hours × 50% = 10 hours
```

or

```
20 hours × 80% = 16 hours
```

In our case, we want to discharge our lead-acid Battery, to the safe 50% depth of discharge, in 5 hours.

So since

```
10 hours × 50% = 5 hours
```

We'll have to look at the C-rate for a 10 hours use, or C_{10}, to know how many Amps we can actually get for our evening consumption.

Let's check the datasheet for one Battery:

Time of discharge	20 hr	10 hr	5 hr
Capacity	100 Ah	92 Ah	78 Ah
Discharge rate	5 A	9.2 A	15.6 A

What we learn here is that if the discharge time is shorter, we get more Amps, but the actual Capacity shrinks a bit.

That's because we'll be drawing much more Current in a shorter amount of time, and this will impact temperature inside the Battery, thus affecting chemical processes and all the stuff we've already covered in the Battery section.

So, looking at the 10 hours column, we know that we can only safely draw 9.2 Amps from each Battery.

```
9.2 Amps × 3 lead-acid Batteries = 27.6 Ah
```

That's still far from the 47.1 Ah we need in the peak hour, but that's more than enough to power the remaining four hours, which will have a maximum Current need of 17.1 Ah.

What we can do in this situation is instead grow our Battery bank to match our needs (6 will do):

```
9.2 Amps × 6 lead-acid Batteries = 55.2 Ah
```

Or reduce our consumption. For example, we can decide to switch off half the lights and avoid charging our devices while someone is taking a shower. That's what I'm going to do here.

AC loads	Wattage	Peak hour	Watt-hours for Peak hour
Hair dryer	1000 W	0.25	× 1 units = 250 Wh/d
			TOTAL (AC) = 250 Wh/d
DC loads			
Water pump	40 W	1	× 1 units = 40 Wh/d
Fridge	45 W	1	× 1 units = 45 Wh/d
Bath fan	25 W	1	× 1 units = 25 Wh/d
Lights (3)	19.2 W	1	× 3 units = 576 Wh/d
		167,6	TOTAL (DC) = 167.6 Wh/d
GRAND TOTAL			(AC+DC) = 417.6 Wh/d

```
417.6 Wh ÷ 12 V = 34.8 Ah
```

That's still more than the 27.6 Ah available, so you still have to buy one more 100 Ah Battery:

```
9.2 Amps × 4 lead-acid Batteries = 36.8 Ah
```

Or you can choose to cut on the hair dryer, which is a huge consumer.

That's it! Now we know how to properly size our Battery bank.

Planning for Critical Situation

One last thing about Battery Banks. If you plan to travel far from any grid, you may want to avoid finding yourself without any electricity. That can happen if you get some rainy days, or if for some reason, the Solar Panels can't produce enough energy to sustain your needs for a few days in a row.

In this scenario, you may want to build a bigger Battery Bank to ensure you never run out of Power.

What I suggest is to triple the Battery Bank size in regards to your peak hour needs.

Or, if you are a super-planner like me, you may want to prepare a backup plan with a consumption reduction schedule based on some scenarios from best to worst case.

That's up to you!

To sum it up:

- Vehicle to power: RV
- System Voltage: 12 V
- Average Electricity daily need: 1,431 W (119.25 Ah)
- Peak Electricity hourly need: 417.6 Wh (34.8 Ah)

- Battery bank: 4 × 100 Ah lead-acid Batteries (9.2 Amps each at C_{10}, or 36.8 Amp-hours total at C_{10})

Calculating Solar Power Output

Once you've sized your needs, and Battery bank, this step is fairly easy.

We've got a 12 V system, so we'll go for 17 V Solar Panels, which are the most common ones.

We need at least 1,431 W per day to serve our needs, but that's not the figure we want to look for.

Instead, we want to consider the Wattage needed to top-up our Battery bank, which is composed of four 100 Ah, 12 V Batteries.

```
Watts = Volts × Amps
12 V × 400 Amps = 3,600 W
```

Since we won't be discharging over 50%, we'll actually need half of that.

```
3,600 W ÷ 2 = 1,800 W
```

Hooray! We just need to buy some panels to get to 1,800 Watts and we're done, right?

I want 200 W panels, so that makes:

```
1,800 W + 200 = 9 Solar Panels
```

Whaaaaat?! Really?

In fact, no. Forget about that.

~~1,800 W + 200 = 9 Solar Panels~~

A Solar Panel's output is calculated on the peak sun hour. That means that a 200 W Solar Panel will only produce 200 W in a peak sun hour.

Therefore, you'd want to check how many peak sun hours your area receives. Note that peak sun hours are different from sunlight hours. They are, in fact, the most intense sunlight hours of the day.

We may use an example here. Let's check the peak sun hours in Nashville, TN.

- High season: 5.2
- Low season: 3.14
- Average: 4.45

We want to take the low season figure, since we want our system to be prepared for the worst-case scenario.

So, our 200 W Solar Panel can produce 200 W in a single peak sun hour. It happens we have 3.14!

Let's do the math:

```
200 W × 3.14 peak sun hours = 628 W per day
```

Of course, PV modules will still produce energy in the remaining average 7 hours of daylight, but we can't rely on imaginary sunny days, weeks, or month. That's why we are only going to consider peak sun hours.

Ok! So we know that one 200 W Solar Panel will give us 628 W of Power every day. Now we need 1,800 W.

```
1,800 W + 628 W = 2.86 Solar Panels
```

That will make three 200 W Solar Panels to fulfill our energy needs. That's way less than nine!

I don't know about yours, but my bank account feels better now.

Planning the locations

Solar

Since every situation is different, here's a bunch of rules-of-thumb you may want to follow.

First, be sure to check how much weight the roof can hold.

Then, take the time to measure the actual free space available to install PV modules. This is a critical factor when it comes to choosing the size, type, and Wattage of Solar Panels. With less space, you'll need to concentrate more Wattage in smaller panels, leading to choose more efficient solutions.

Remember to consider the kind of installation you want: flat? With a tilt mount? Etc. This will have a significant impact on the weight and the space needed.

If you are building a Solar power system for a boat, consider the sails and the shadow they cast.

Battery Bank

You'll need a sweet spot to store your Batteries. This spot should be easily accessible for you to install, do the maintenance stuff, and bring down the cables from the charge controller, the inverter, and the inside of the Rv/cabin/boat.

You'll have to make room for all the Batteries, possibly considering future integrations.

You'll have to give them room to breathe (above, below, and all around) since we know the temperature has a significant impact on their performance and lifespan.

You'll have to ensure the area is or can be ventilated often, but also make sure water will never ever get in.

Everything Else

The Charge Controller and Inverter should be placed as close to the Battery bank as possible.

As a general rule-of-thumb, try to minimize the wire length. It reduces energy losses and makes everything less messy.

Choosing the Charge Controller and the Inverter

Charge Controller

Apart from efficiency and overall performance, which will depend on your budget, there are only two things to consider when choosing a Charge Controller.

First, Voltage. Just make sure the Charge Controller is able to deal with the Voltage coming out of the Solar Panels. In our case, it's 17 Volts, so a 24V Charge Controller will be ok.

Second, wiring type. If you wire your panels in parallel, you can buy either PWM or MPPT. But if you wire your panels in series, then MPPT is compulsory.

Inverter

Inverters take a DC signal from the Battery bank and turn it into an AC signal to power your favorite appliances.

Most of the Inverters can give you 120/240 V output, which is more than enough in almost any case.

What you want to check here is the Wattage, the Inverter can handle since you have to match it to your consumption.

Practically, you need to check the Inverter's peak draw and match the peak hour consumption you're going to have. Then check the continuous draw of the Inverter, and match the average Power consumption planned.

In our case, we have an esteemed peak consumption of 417.6 Wh.

As per the continuous consumption, we've previously made the math and got 17.1 Ah:

```
17.1 Ah × 12 V = 205.2 Wh
```

Frankly speaking, that is actually a sort of peak-continuous consumption since we are sizing everything around the most power-consuming time of the day, which is the evening.

In fact, the daily Watt-hours consumption will be lower than that.

Let's wrap it up.

We need an Inverter capable of:

At least 417.6 Wh peak draw.

At least 205.2 Wh continuous draw.

Of course, the more Watts your Inverter can handle, the safer. Since you'll never have to worry about excessive Power draw on the AC side.

Also, check manufacturer specs for the single plugs coming out of the Inverter since they may not handle the whole of the Power the Inverter is rated for.

Sizing Wires

Where applicable, we'll be using stranded wire since it is far more resistive for our on-the-road use. To calculate the right wire size, we have to estimate the maximum Current that will be flowing through the specific section of our system.

To do so, we'll be using our dear old formula:

```
Amps = Watts ÷ Volts
```

Once we know how much Current Capacity we are going to need our wires to provide, we'll just have to check the AWG chart.

Size (AWG or kcmil)	Temperature Rating of Copper Conductor					
	60°C (140°F)	75°C (167°F)		90°C (194°F)		
	Types TW UF	Types RHW XHHW THHW USE THW ZW THWN		Types FEP SIS USE-2 FEPB TBS XHH MI THHN XHHW RHH THHW XHHW-2 RHW-2 THW-2 ZW-2 SA THWN-2		
18 AWG	—	—		14		
16 AWG	—	—		18		
14 AWG*	15	20		25		
12 AWG*	20	25		30		
10 AWG*	30	35		40		
8 AWG	40	50		55		
6 AWG	55	65		75		
4 AWG	70	85		95		
3 AWG	85	100		115		
2 AWG	95	115		130		
1 AWG	110	130		145		
1/0 AWG	125	150		170		
2/0 AWG	145	175		195		
3/0 AWG	165	200		225		
4/0 AWG	195	230		260		
250 KCMIL	215	255		290		
300 KCMIL	240	285		320		
350 KCMIL	260	310		350		
400 KCMIL	280	335		380		
500 KCMIL	320	380		430		
600 KCMIL	350	420		475		
700 KCMIL	385	460		520		
750 KCMIL	400	475		535		
800 KCMIL	410	490		555		
900 KCMIL	435	520		585		
1000 KCMIL	455	545		615		
1250 KCMIL	495	590		665		
1500 KCMIL	525	625		705		
1750 KCMIL	545	650		735		
2000 KCMIL	555	665		750		

* Unless otherwise specifically permitted elsewhere in the NEC NFPA70 Code, the overcurrent protection for conductor types marked with an asterisk shall not exceed 15A for No. 14 copper, 20A for No. 12 copper, and 30A for No. 10 copper, after any correction factors for ambient temperature and number of conductors have been applied.

Solar Panels to Combiner box

The wires here are usually provided by the Solar Panel manufacturer, but let's do the math anyway.

```
Voltage: 17.6 V (PV module)

Wattage:  200 W (single PV module)

Amps = 200 ÷ 17.6 = 11.3 A
```

We are going to use a RHW-2 copper wire, so by looking at the AWG chart, we know that we need a 18 AWG wire.

Combiner Box to Charge Controller

From the combiner box, all the Current from our 3 200W 17.6 V panels will be carried to the Charge Controller.

```
Voltage: 17.6 V (PV module)

Wattage:  200 W × 3 solar panels = 600 W

Amps = 600 ÷ 17.6 = 34.1 A
```

We are going to use a THHN-2 copper wire, so by looking at the AWG chart, we know that we need a 10 AWG wire. But, as we can see at the bottom of the chart, 10 AWG wires can actually only safely handle 30 A. Our final choice should therefore be an 8 AWG wire.

Charge Controller to Batteries

The Charge Controller will modulate the output Voltage to match our Batteries' Voltage, that is 12 V.

```
Voltage: 12 V

Wattage:  600 W

Amps = 600 ÷ 12 = 50 A
```

We are going to use a THHN-2 copper wire, so by looking at the AWG chart, we know that we need an 8 AWG wire.

Batteries to Batteries to Inverter

Talking about Batteries, that's where our system will make the most effort, since it's from there that the appliances, through the Inverter, will draw Power.

We already calculated that our peak hour consumption will be 36.8 Ah.

We've built our Battery bank around that rating, so we really shouldn't be even considering exceeding that.

But life is life, and for extra wire-safety, what we can do is sizing the wire diameter taking into account the maximum Current possibly drawn in case we use all the appliances together at the same time. We already calculated that too, and it makes 119.25 Ah.

So, two options here.

Either we go for an 8 AWG THHN-2 copper wire, able to deal with the peak hour 36.8 Ah.

Either we go for a 2 AWG THHN-2 copper wire, able to deal with the peak hour 119.25 Ah.

Batteries to Inverter

The Inverter only powers AC appliances, therefore we will size the wire according to a scenario where all AC appliances are being used at the same time. We already calculated that, and it's 610 Wh.

```
Voltage: 12 V
Wattage:  600 W
Amps = 610 ÷ 12 = 50.83 A
```

We are still going to use a THHN-2 copper wire, so by looking at the AWG chart, we know that we need an 8 AWG wire.

Inverter to Loads

Most Inverters have plugs to connect your appliances directly, so you won't need wires.

If it happens that you needed wires anyway, check the appliance Current demand, and size the wire accordingly.

Batteries to DC Loads

It depends.

If you wire each DC appliance directly to the Battery, you'll need to size each wire according to the appliance Current demand.

If you wire each DC appliance to a Busbar, then the same applies to each single DC connection, but you'll also have to connect the Battery to the Busbar.

To wire the Battery to the Busbar, you'll need to calculate the power demand of the DC loads as if they were being used all together at the same time. We've already calculated this too, and it makes: 821 Wh.

```
Voltage: 12 V
Wattage:  821 W
Amps = 821 ÷ 12 = 68.42 A
```

Therefore, as we'll still be using a THHN-2 copper wire, we know that we need an 6 AWG wire.

Sizing Fuses

Fusing is meant to protect the components in the circuit and the circuit itself. An excessive amount of Current can cause the wires to burn out and start a fire.

Each fuse should be sized based on the Amperage of the wire (always a positive red line) they will be put on.

Fuses never have to exceed 1.5 times (or 150%) of a wire's Amperage. For safety reasons, this factor is usually brought down to 1.25 times (or 125%).

If you would install a Fuse that has a lower Amperage than the wire (and the loads connected to it, since you should have based the wire size on the load needs), the fuse will melt as soon as you turn the system on.

Let's bring on an example from our system here. Let's size the fuse we will need between the Charge Controller and the Battery Bank.

As calculated earlier, we have 50 A.

```
50 A × 1.25 = 62.5 A
```

Therefore, we will be using a 63 A fuse.

More often than not, Solar Panels will have specific ratings for fusing, so be sure to check the datasheet for the maximum allowed fuse each PV module can bear.

Also, note that upstream fuses need to be bigger than downstream fuses. Upstream here is identified with the power source, which can be PV modules when they are running, or Batteries at night.

Fuse Box

In some cases, you might prefer fuse boxes instead of single fuses. A fuse box makes for the tidiest environment and may be considered for DC loads to avoid ending up tangled.

Notes on Appliance Efficiency

Most of our AC appliances, such as laptops and smartphones, are actually DC appliances since they run on electronics.

What is AC is not the appliance itself, but rather the charger. Chargers are designed to take the Current from standard AC home wall plugs and convert this signal to DC. Therefore, every charger is, in fact, an Inverter.

But the process of "inverting" electrical signals consumes energy. Thus it is not very efficient to have electricity from a DC power source (Battery) that is converted to AC by the Inverter (energy loss), and then having a laptop charger that inverts the signal back to DC (second energy loss) to operate the device.

What you can do to reduce these energy losses is to buy DC chargers for your AC (but actually DC) devices. Note that car chargers for smartphones are already DC.

Many Inverters have a DC output plug, so it will suffice to plug your new DC charger into that, and voilà!

If your Inverter doesn't come with a DC plug, though, you'll need to build your own DC plugs and wire them directly to the Battery, or the DC Busbar. Note that USB plugs are DC plugs, so you may want to build USB plugs around your RV, boat, or cabin.

Of course, the DC port and charger should match the Voltage coming out of the Battery bank.

Final System Design

So, we now have everything we need to design our system. We'll be using Busbars to have a tidier design between Batteries and Loads. Let's lay this down.

Solar to Combiner box (parallel wiring)

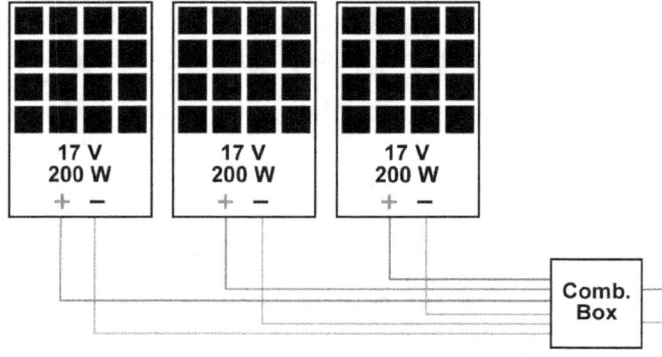

Combiner box to Charge Controller

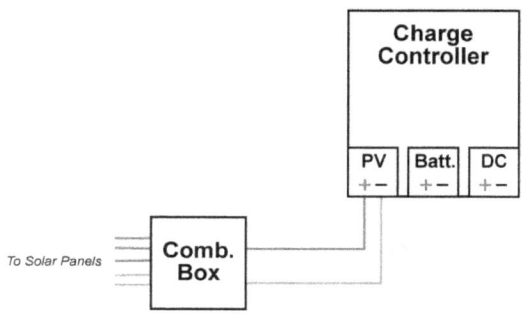

Charge Controller to Batteries, and Batteries to Batteries (parallel wiring)

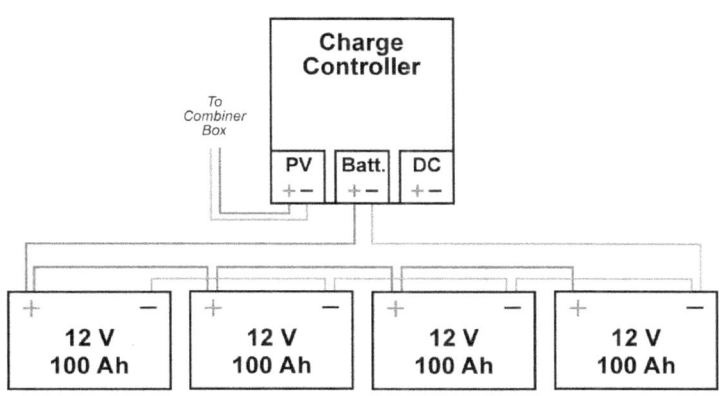

Batteries to Busbars

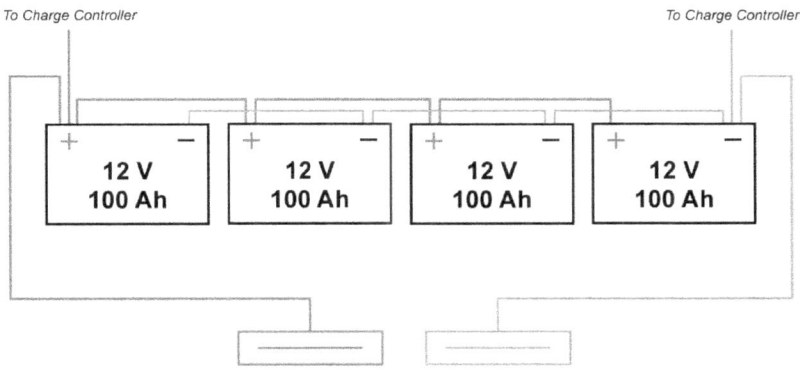

Busbars to Inverter

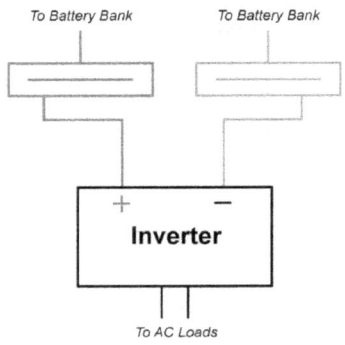

Busbars to DC Loads

With single fuses

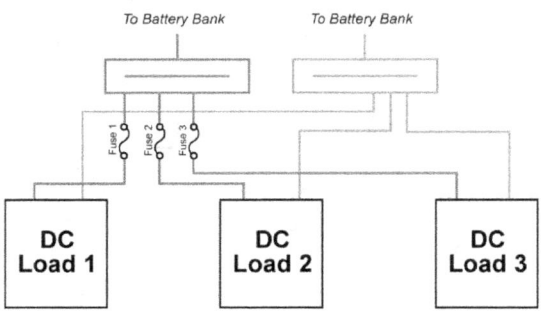

With a Fuse box

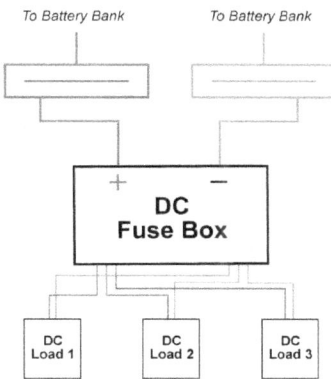

Safety: Fuses and Switches

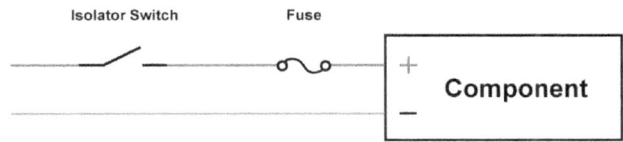

Hey Andy? Why so many Switches? Well, because you'll possibly want to do some maintenance to a part of the system without having to shut down everything. With this setup, we can be using AC appliances while repairing DC loads, etc.

Optional: Shunt and Battery Monitor

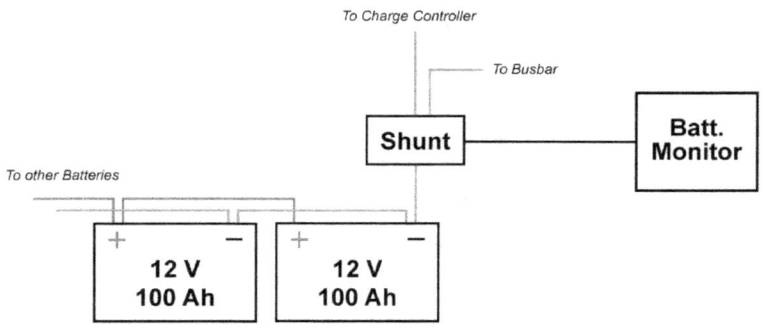

Final System Blueprint

Without Shunt

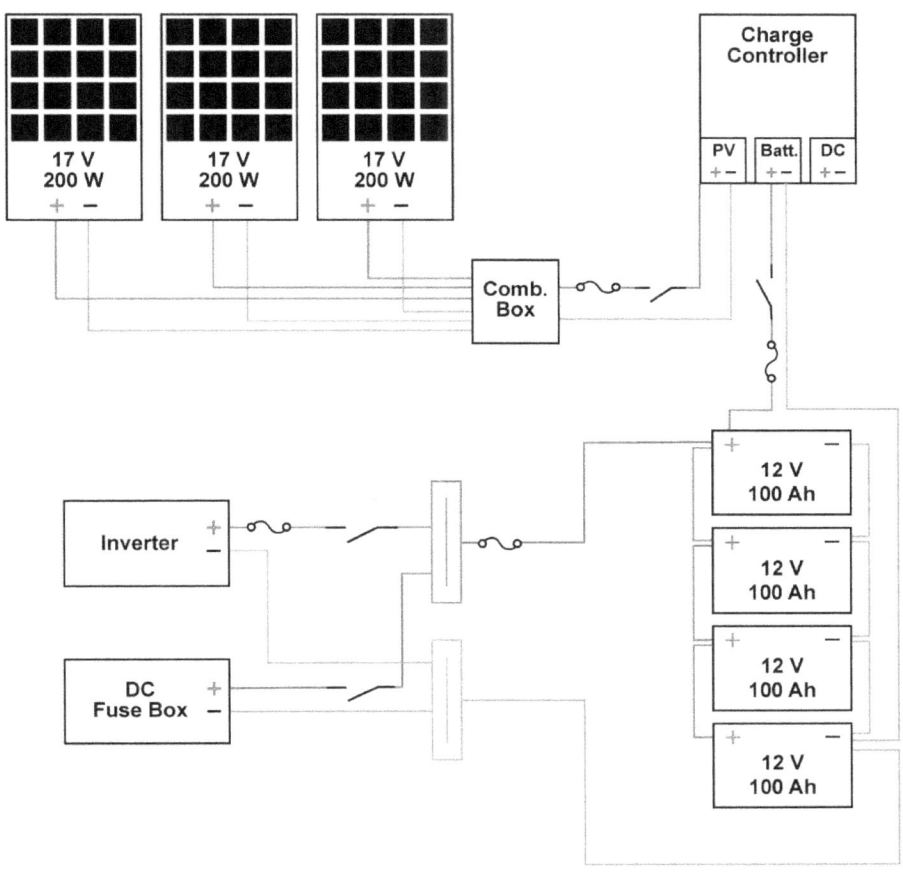

Final System Blueprint

With Shunt

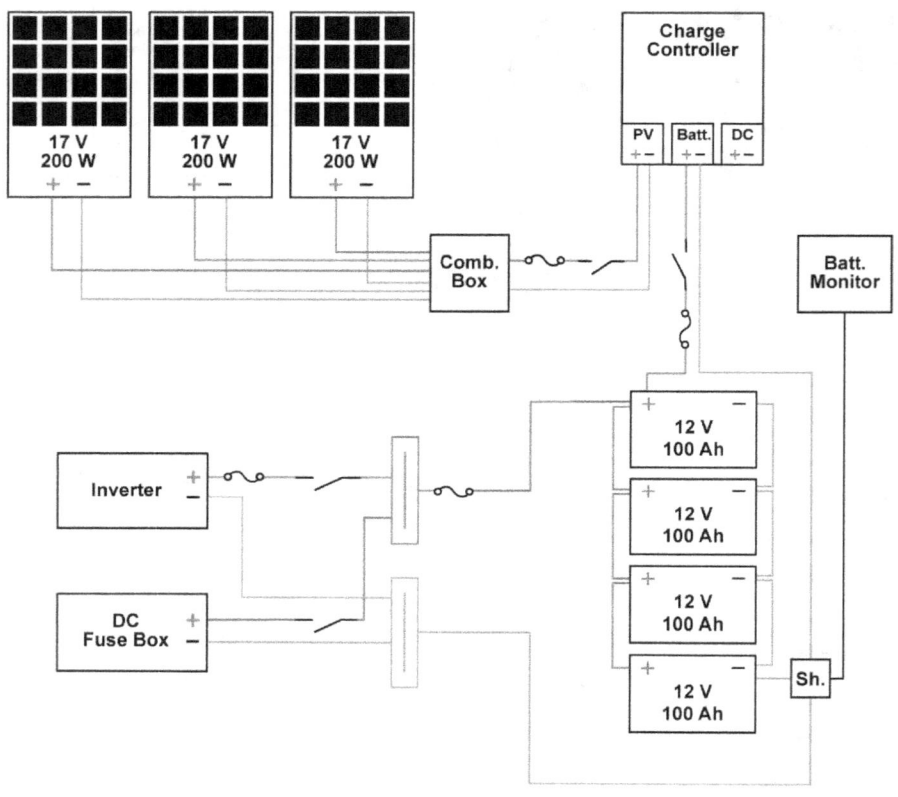

Setup

Now that we have all our components and a properly designed blueprint for our system, it's time to make things happen!

These are the tools you will need and the procedure you'll need to follow.

Let's get our hands dirty!

Tools and Equipment

Gloves

Remember, electricity can be hazardous. Wearing electricity resistant gloves will protect you against shocks. Many think they can do without insulating rubber gloves, leather protectors, etc. Don't make a mistake, don't save money on safety.

Safety Goggles

Always protect your eyes when working with tools, drilling holes, and playing with electricity.

Adhesive

In boats and RVs application, you may not want to drill your vehicle's roof or shell and risk water to pour down that hole.

Super strong adhesives exist and work pretty well, even in harsh conditions.

Drill

Whether you are going to wall components or secure them on the floor, you'll need a drill.

Hacksaw and Holesaw

You will most likely be needing to cut some holes to allow your wires to pass from a compartment to another. Thus, a holesaw can be even more useful than a simple drill.

Hacksaws cannot cut holes, but you're gonna need them at one time, trust me.

Level

You will need a level to install any device or component properly. It will ensure you are walling components balanced and straight.

Screwdrivers, Wrench and Ratchet Set

You will need them to mount the components or tighten the cables to terminals, such as Battery's. Look for screwdrivers that have been tested to sustain a fair amount of Voltage in both DC and AC.

Crimping Tool

It will allow for securing wire connectors and crimp terminal for different AWGs.

Conduit Cutter

If you use conduit to protect your cables, which will most likely be the case for the exposed Solar wires on the roof, a conduit cutter will be needed. Of course, if you use a metal conduit, be sure to buy a metal conduit cutter.

Wire or Cable Cutter

How are you going to wire anything without a wire cutter? For most heavy works, you might need a cable cutter, though.

Pointy-Nose Pliers

You will need them to bend, re-position and snip the wires.

Wire Stripper

It will help to strip and cut wires made of copper or aluminum without damaging them.

For cables from 5AWG to 4/0 AWG, you will need a cable stripper, though.

Hammer Lug Crimper

Especially useful for Battery banks, it will help you crimp a wide variety of cable lugs.

Wire Tape

Also known as insulating tape, or electrical tape, wire tape is a pressure-sensitive tape, usually made of vinyl, that allows you to insulate electrical wires and other conductive materials. It can be useful to secure some connections, repair some cables, and even labeling them.

Setup Procedure

When you have a proper blueprint of the circuit printed out.

When your components are all lined up.

When the locations are identified.

When you've done the drilling of holes and stuff.

Then you're ready to set up your 12 V Solar Power system!

IMPORTANT: Remember to use protection while working on the circuit, and remember to keep each Isolator Switch you install turned off.

1. Batteries

1. **Placement:** Place the Batteries in the chosen compartment.

2. **Parallel wiring:** to wire them in parallel, take an appropriately sized red wire, and connect the positive terminals of the different Batteries. Then use the black wire to connect the negative terminals of the different Batteries together as well.

3. **Fuse and Isolator Switch:** Fuse and Isolator Switch will go on the positive wire (red) coming down from the Solar Charge

Controller to the first Battery of the bank. The Fuse should be placed as close as possible to the Battery.

4. **Shunt:** if you are going to use a Shunt, now is the time to wire it. The Shunt will go on the negative wire (black) coming out of the last Battery in the bank. Like the Fuse, connect it as close as possible to the Batteries. Note that the negative terminal of the Shunt will act as the negative terminal of the Battery.

2. *Charge Controller*

1. **Input terminals:** Charge Controllers have two input terminals, one for PV modules and one for Batteries. They also have an output terminal for DC current, that we will ignore here.

2. **Connecting the positive wire:** take the positive (red) wire coming from the first Battery. That should now also pass from a fuse and an Isolator Switch and connect it to the positive Battery terminal on the Charge Controller.

3. **Connecting the negative wire:** now take the negative (black) wire coming from the last Battery in the series, and connect it to the negative Battery terminal on the Charge Controller. If you have a Shunt installed, the negative wire

going to the Charge Controller will come from the Shunt's negative terminal.

4. **Setting up the Charge Controller:** once everything is connected, you can close the circuit to allow electricity to run through the wire and get to the Charge Controller. The display will light up. From now on, follow the instructions in the manufacturer's manual. When you are done, remember to turn off the Isolator Switch and break the circuit.

3. Battery Monitor

If you chose to use a Battery Monitor (wise choice), the cable coming with the Battery Monitor will have to be connected to the Shunt directly. Follow the manufacturer's instructions for more detailed guidance.

4. Inverter

When it comes to wiring the Inverter, you have two options available.

1. **Option 1: wiring directly from the Battery.**

    a. **Fuse and Isolator Switch:** Take the positive (red) wire from the first Battery in the series, and connect it to an Isolator Switch. Then connect the Isolator Switch to a Fuse appropriately sized, placed as close as possible to the Inverter.

    b. **Wiring the Inverter:** take the positive (red) wire from the Fuse, and connect it to the positive terminal of the Inverter. Then take the negative (black) wire from the last Battery of the bank, or the Shunt if you have one, and connect it to the Inverter.

2. **Option 2: using two Busbars.**

    a. **Fuse:** take the positive (red) wire from the first Battery in the series, and connect it to a Fuse appropriately sized, placed as close as possible to the first Busbar. We won't be using Isolator Switches here, because we are going to use them on the other side of the Busbar to isolate AC and DC loads separately.

    b. **Wiring the first Busbar:** take the positive (red) wire from the Fuse, and connect it to the first Busbar.

c. **Wiring the second Busbar:** take the negative (black) wire from the last Battery in the bank, or the Shunt if you have one, and connect it to the second Busbar.

d. **Wiring the Inverter and Isolator Switch:** connect one side of a red wire to the first Busbar, and the other to an Isolator Switch. Then connect another positive wire from the Isolator Switch to a carefully sized Fuse, and finally wire the Fuse to the positive terminal of the Inverter. Then connect one side of a black wire to the second Busbar and the other to the Inverter's negative terminal.

5. *DC Loads*

Again, we have two options here. Since we have multiple DC loads in both cases, we will be using a DC Fuse Box.

1. **Option 1: without Busbars.**

    a. **Positive to DC Fuse Box and Isolator Switch:** wire another positive (red) wire to the positive terminal of the first Battery in the bank, and connect the other end to an Isolator Switch. Then connect the Isolator Switch to the DC Fuse Box.

b. **Negative to DC Fuse Box:** wire another negative (black) wire to the negative terminal of the last Battery in the bank, or the Shunt if you have one installed, and connect the other end directly to the DC Fuse Box input DC negative terminal.

c. **DC Fuse Box to DC load:** in a Fuse Box, each fuse is sized for a specific load's Current need (make sure that it is). So you'll have to wire a positive (red) wire from the single specific Fuse to a load's positive terminal. Then, wire one of the many output negative terminals of the DC Fuse Box to the negative terminal of the DC load.

2. **Option 2: with Busbars.**

a. **Positive to DC Fuse Box and Isolator Switch:** wire a positive (red) wire to the positive Busbar, then connect the other end to an Isolator Switch. Then wire a positive (red) wire from the Isolator Switch to the positive terminal of the DC Fuse Box.

b. **Negative to DC Fuse Box:** wire a negative (black) wire to the negative Busbar, then connect the other end to the negative input terminal of the DC Fuse Box.

c. **DC Fuse Box to DC load:** in a Fuse Box, each fuse is sized for a specific load's Current need (make sure that it is). So you'll have to wire a positive (red) wire from the single specific Fuse to a load's positive terminal. Then, wire one of the many output negative terminals of the DC Fuse Box to the negative terminal of the DC load.

Note on Fuse Boxes: most Fuse Boxes already have a single bigger fuse built-in to protect itself. If it wasn't the case for your specific Fuse Box, you'd have to place a Fuse right before the Fuse Box on the positive (red) wire.

6. Solar Panels

1. **Install the Panels:** secure them to your vehicle or cabin, drill the roof, or use high strength adhesive.

2. **Parallel wiring:** to wire Solar Panels in parallel, take the positive (red) wire of each individual module and bring them to the Combiner Box. Inside the Combiner box, you will find one positive (red) Busbar and one negative Busbar. Wire the positive wires to the positive Busbar. When you're done, do the same with the negative wires coming from the PV modules.

3. **Combiner Box to Fuse, Isolator Switch and Charge Controller:** now, you will have one single beautiful positive (red) wire and one single beautiful negative (black) wire coming from the Combiner Box. Connect the positive (red) wire to an appropriately sized Fuse, and then to an Isolator Switch, following the same procedure we referred to until now. Then connect the negative (black) wire coming from the Combiner Box, and the positive (red) wire coming from the Isolator Switch to the negative and positive PV input terminals on the Charge Controller, and you're done.

We won't need to Fuse each PV module separately here, since most of the time, single Fuses are already integrated into the Solar Panels. However, if fuses are not integrated already, I suggest to fuse each positive line coming from the Panels.

Before connecting Solar Panels wires to the Charge Controller, remember to ground them, as explained in the specific section earlier in this book.

7. *Testing*

Well done! You have now completed your first Solar Power System installation!

Before enjoying your sun-harvested electricity, it is essential to run some tests to make sure everything is up and running.

1. Check for loose wires.

2. Look for sharp edges that can menace your cables.

3. Check the temperature of each component.

4. Also, check the temperature of the wires.

5. Monitor Battery Voltage when your bank is fully charged.

6. Test the functioning of each load.

7. Look for possible errors in the setup by double-checking your circuit blueprint.

8. Labeling

I strongly recommend taping each and every wire to quickly grasp, with a glimpse of an eye, what the cable is about. This will make it easier to do the maintenance, and provide a neat working environment for whoever will be dealing with the system in the future.

Maintenance and Storage

Installing a Solar Power system can be quite expensive in the first place, and you want to make components last as long as possible.

That's why you need to know how to keep everything up and running efficiently, and how to treat components to extend their lifespan.

Cleaning Solar Panels

Maintenance is key to make sure your Solar Panels keep producing energy. Since dirt gets in the way of sun rays, it affects performance severely. Therefore you'll need to keep the PV modules as clean as possible.

Ok, Andy, just tell me how to do it.

Right! First of all, you have to disconnect the Panels by turning off the isolator switch.

Then you have to get some de-ionized water. Avoid regular water since its minerals will stick to the glass, and that is the opposite of what we want.

Submerge a sponge in the de-ionized water, which has to be at room temperature, and start cleaning until any visible dirt is gone.

PV modules can get very hot, so it is best to do the cleaning early in the morning or late in the afternoon. You also want to avoid getting in the way in the peak production hours.

Storing Batteries

Batteries need to be taken care of at least every six months, especially lead-acid ones. They should be completely clean, without any corrosion that can cause contact issues, and loosen the connections.

First thing is to make sure the Battery is 100% charged by checking the Charge Controller, the Battery Monitor, or by using a voltmeter (but that should be done in an open circuit, without any load attached).

Some safety measures now.

Isolate the Battery by turning off the switches to the Inverter, Charge Controller, and DC loads, or Busbar. I suggest you just switch everything off, including Solar Panels, and the rest.

Wear protective gloves, goggles, and also long sleeves and pants, if you can.

Also, make sure to place the Battery on a non-conductive surface.

As for the Solar Panels, use a sponge and de-ionized water. You can use distilled water as well, but de-ionized is non-conductive, which is best. If needed, you can mix the water with sodium bicarbonate (100 g, or 3.5 oz per liter).

Clean everything up, including terminals and clamps, to remove dirt, dust, corrosion, etc. Loose connections can cause sparks and possibly start fires. And we don't want that. Make sure everything is dry before putting them back in place.

When unused, Batteries should be switched off the circuit and stored in a safe place. The temperature of the storage room should be between 5°F (-15°C) and 122°F (+50°C), but be sure to check the manufacturer's information for your specific Battery.

The Battery will self-discharge even if they are not used. It is best to charge them up every month. For a more precise schedule, it suffices to look for the self-discharge rate on the datasheet, and do the math. Always keep in mind the actual Depth of Discharge, as we discussed earlier.

Notes on Maintenance

Note 1: When doing maintenance, always turn the isolator switches off, and make sure appliances are turned off as well. If they are not, they will be as soon as you turn the isolator switches off, but it is preferable to singularly turn them off before that.

Note 2: Always protect yourself with working equipment such as gloves and goggles.

Note 3: Always check the datasheet of the specific component for precise information from the manufacturer on how to do the maintenance.

Conclusion

Congratulations my friend! You've reached the end of this book. But I guess it was more of a personal journey that led you here. The inmost desire to build an off-grid system that means energetic self-sufficiency.

Go back, for a moment, to the very beginning of this journey. To the day you bought this book, prepared to undertake this voyage into the solar realm and its wonders.

Now, look at you. You've made it. You've learned a lot. You've accomplished so many things. And you are now the proud owner and maker of a splendid DIY 12 V Solar system built from scratch. You've taken a dream and made it happen. Who could be stopping you now?

Now you can enjoy real freedom. You have freed yourself from the bills (eh eh) and from the grid. You can now roam free wherever you want with your boat, van, RV, or whatever.

And you can also be proud to contribute to clean energy production and reduction in fossil fuel burning.

Congratulations again. I hope one day we could meet somewhere on the road. And talk about the fascinating wonders we've discovered since Solar Power entered our lives.

See you soon!

Andy

Will You Do Me a Favor?

Hello friend, glad you're here!

Before you go, can I ask you a small favor?

Would you take a minute or two and write a comment about this book on Amazon?

This is an independently published book, and I have no other way to get feedback and make it better.

Copy the following link in the URL bar of your browser to leave a comment on Amazon.com:

https://bit.ly/OffGridSolarPowerHandbook

I've put so much effort into this book, and it would mean a lot to me. Thank you so much, my friend.

If the above link doesn't work, these things are painful sometimes, please navigate to the book page manually. Double thank you if this is the case!

Andy

www.ingramcontent.com/pod-product-compliance
Lightning Source LLC
Chambersburg PA
CBHW060836220526
45466CB00003B/1128